Acknowledgments
鸣谢

The research of this studio would not have been possible without the generous support of the Harvard China Fund, which sponsored the travel of GSD professors and students to Beijing during February of 2011. Among the profound and rich experiences afforded by this funding were collaboration with professors, experts, and students from Peking University; meetings with Chinese planning and municipal officials; and access to the region and people of Taihuzhen. We are grateful to the Mayor of Taihu Township, Mr. Ni Decai, who had the initial inspiration to seek out alternative sustainable futures for his town; to Professors Wu Honglin and Kelly Shannon and their PKU students for their informative ideas and warm welcome; and to the office of Turenscape, which provided invaluable logistical and material support. At the GSD, Dean Mohsen Mostafavi and the departments of Landscape Architecture and Urban Planning and Design all provided essential encouragement and support; Professor Richard Forman advised us on ecological frameworks in urban contexts; Dr. Betsy Colburn spoke with us about hydrology; and doctoral candidate Jorge Colon explained nuances of informal settlement in Chinese cities. We thank Professors Joan Busquets, Alex Krieger, Rahul Mehrotra, Felipe Correa, Rafi Segal, Chris Reed, Pierre Belanger, Jane Hutton, Adrian Blackwell, and architect Douglas Voigt for the expert insights they brought to our midterm and final reviews.

感谢哈佛大学中国基金会对本次设计课程研究的慷慨支持和对2011年2月哈佛大学设计学院师生北京之行的资助。通过与北京大学的教授、专家和学生的合作，与中国的规划和市政官员的会谈以及台湖镇人民的帮助，我们得到了他们深刻而丰富的经验。我们感谢台湖镇镇长倪德才，他首先提出寻找台湖镇可持续未来多解方案的想法；感谢武弘麟教授、凯利·香农教授以及北京大学学生的建设性想法和热情欢迎，感谢土人景观宝贵的支持和帮助。感谢哈佛设计学研究院院长莫森·莫斯塔法维及景观设计学系、城市规划设计学系给予的鼓励和支持，感谢理查德·福尔曼教授关于城市布局中生态框架的建议，感谢贝齐·科尔伯恩博士关于水文状况的分析，感谢博士候选人乔治·科隆关于中国城市棚户区的差异分析。还要感谢琼·布斯克茨教授、亚历克斯·克里格教授、拉胡·梅罗特拉教授、费利佩·科雷亚教授、拉菲·西格尔教授、克里斯·里德教授、皮埃尔·贝朗格教授、简·赫顿教授、阿德里安·克莱克韦尔教授以及建筑师道格拉斯·沃伊特在中期和最终评审时提出的专业意见。感谢《景观设计学》编辑部对本书的校对。

Image Credits
Page 24, Bottom Left: Beijing Institute of City Planning and Design
Page 24, Bottom Right: Beijing Institute of City Planning and Design
Page 32, Bottom Left: Beijing Atlas, 1994
Page 34, Bottom Left: Beijing Institute of City Planning and Design
Page 36, Bottom Left: Beijing Institute of City Planning and Design
Page 47, Bottom: Beijing Municipal City Master Plan

图片来源：
24页，左下：北京市城市规划设计研究院
24页，右下：北京市城市规划设计研究院
32页，左下：1994，北京地图
34页，左下：北京市城市规划设计研究院
36页，左下：北京市城市规划设计研究院
47页，下：北京市总体规划

All photos, unless noted otherwise, taken by the following:
Cameron Barradale, MLAUD
Andreas Viglakis, MAUD
Kongjian Yu, Professor of Landscape Architecture (Peking University)
Hyung Seung Min, MAUD
Troy Vaughn, MLA
Mark Mulligan, Adjunct Associate Professor of Architecture

所有未注明的照片均来自：
卡梅伦·巴拉戴，景观设计与城市设计师
安德烈亚斯·威格拉基，建筑与城市设计师
俞孔坚，景观设计学教授（北京大学）
闵玄胜，建筑与城市设计师
特洛伊·沃恩，景观设计师
马克·马利根，建筑学副教授

The Harvard University Graduate School of Design is a leading center for education, information, and technical expertise on the built environment. Its departments of Architecture, Landscape Architecture, and Urban Planning and Design offer masters and doctoral degree programs and provide the foundation for Advanced Studies and Executive Education programs.

哈佛大学设计学院在环境研究领域的各个方面，包括教育、信息和技术等均有着领袖地位。它的建筑系、景观设计学系和城市规划设计学系均可授予硕士和博士学位，同时提供高级进修和行政教育课程。

Copyright © 2011, The President and Fellows of Harvard College. All rights are reserved. No part may be reproduced without permission.

版权©2011，所有解释权归哈佛学院主席及其成员所有，未经允许，不得翻印。

Graduate School of Design Faculty
Kongjian Yu, Professor of Landscape Architecture (Peking University)
Peter Rowe, R. Garbe Professor of Architecture and Urban Design
Stephen Ervin, Lecturer, Director of Information Technology
Mark Mulligan, Adjunct Associate Professor of Architecture
Hailong Liu, Visiting Scholar (Tsinghua University)
Har Ye Kan, DDesS candidate

Graduate School of Design Students
Cameron Barradale, MLAUD
Yenlin Cheng, MLA
Hana Disch, MAUD
Darin Mano, MArch
Fadi Masoud, MLA
Hyung Seung Min, MAUD
Carrie Nielson, MLA/MUP
Ryder Pearce, MUP
Aditya Sawant, MAUD
Mark Upton, MAUD
Troy Vaughn, MLA
Andreas Viglakis, MAUD

Peking University Students
Annelies De Nijs
Dongsong Deng
Yuchen Du
Jingsi Fang
Ying Gui
Yue Hou
Chao Huang
Yuqing Lai
Panpan Li
Ya Li
Yingyun Lu
Yan Mao
Miao Nie
Meilin Pan
Hua Shu
Liqing Song
Jianning Sun
Lu Sun
Shu Sun
Song Wang
Xiaobing Wang
Peichun Wen
Jieling Xu
Ye Yao
Yuanfang Zhan
Jingru Zhang
Ran Zhang
Tingting Zhang
Zheng Zhang

哈佛大学设计学院教师
俞孔坚，景观设计学教授（北京大学）
彼得·罗，建筑和城市设计学教授
斯蒂芬·欧文，讲师，信息技术总监
马克·马利根，建筑学副教授
刘海龙，访问学者(清华大学)
简夏仪,在读设计学博士

哈佛大学设计学院学生
Cameron Barradale, MLAUD
Yenlin Cheng, MLA
Hana Disch, MAUD
Darin Mano, MArch
Fadi Masoud, MLA
Hyung Seung Min, MAUD
Carrie Nielson, MLA/MUP
Ryder Pearce, MUP
Aditya Sawant, MAUD
Mark Upton, MAUD
Troy Vaughn, MLA
Andreas Viglakis, MAUD

北京大学学生
Annelies De Nijs
邓冬松
杜玉成
房静思
桂颖
侯跃
黄超
赖雨青
李盼盼
李雅
卢映云
毛岩
聂淼
潘梅林
舒华
宋丽青
孙建宁
孙璐
孙姝
王松
王小兵
温佩君
许洁舲
姚晔
詹圆方
张静茹
张然
张婷婷
张峥

Contents

4	Acknowledgments
8	Foreword by Joan Busquets
10	Studio Introduction
16	**Beijing Regional Analysis**
18	Context: China's Bohai Region
20	Beijing's Growth and Peri-urban Issues
22	"Green Wedges" in the Beijing Metropolitan Plan
24	Urban Adjacencies: Tongzhou and Yizhuang
26	**Taihu Site Analysis**
28	Land Use in Taihu
30	Hydrology
32	Agriculture
34	Nature Conservation
36	Economy and Demographics
38	Transportation System
40	Settlement Patterns
42	**Future Prospects + Design Proposals**
44	Essay: Web, Band, Network, and Field Operations: Alternatives to the Conventions of Green Belts and Wedges in the Beijing Metropolitan Plan + *by Peter Rowe*
48	Eco-City Taihu *by Ryder Pearce + Mark Upton + Troy Vaughn*
58	Green Network, Green Web *by Yenlin Cheng + Hana Disch + Aditya Sawant*
68	Urban Landscape Connectivity *by Darin Mano + Hyung Seung Min + Carrie Nelson*
78	Nested Scales of Urbanization *by Cameron Barradale + Fadi Masoud + Andreas Viglakis*
88	Collaboration + Site Visit
94	Afterword *by Kongjian Yu*

目录

4 鸣谢

8 前言 琼·布斯克茨

10 课程简介

16 北京区域分析
18 地理背景:中国环渤海地区
20 北京的城市扩张和郊区问题
22 北京市城市总体规划中的"绿楔"
24 城市邻接:通州和亦庄

26 台湖场地分析
28 台湖的土地利用
30 水文
32 农业
34 自然保育
36 经济与人口
38 交通系统
40 居住模式

42 前景展望和设计方案
44 网、带、网络和场地运营:
  北京市城市总体规划中绿带和绿楔的多解方案 彼得·罗
48 台湖生态城 赖德·皮尔斯、马克·厄普顿、特洛伊·沃恩
58 绿色网络 陈燕霖、哈娜·迪施、阿迪亚·萨瓦特
68 城市景观连通性 达林·马诺、闵玄胜、卡莉·尼尔森
78 城市化的叠加尺度 卡梅伦·巴拉戴、法蒂·马苏德、安德烈亚斯·威格拉基

88 合作场地参观

94 后记 俞孔坚

Foreword
by Joan Busquets
前言
琼 · 布斯克茨

This studio examines the urbanization and development of Taihu, a formerly rural area now being transformed by expansion along the Beijing-Tianjin axis – currently taking shape as one of the strongest axes of growth in the urban system formed around China's capital.

The studio's research serves to highlight some of the major contradictions in the frenetic development of such a dynamic urban system as that of Beijing: for example, the contrast between the model of circular ring roads and the buffer that is the second green belt of parks proposed in the city's master plan on the one hand, and the urbanizing force of the radial axis of infrastructures with the high-speed train and the expressways to Tianjin on the other (both the high-speed rail and expressways are recently built and tending to unbalance the "belt model"). Another issue is the difficulty of preserving areas of high agricultural and ecological value in the face of the new economic and urbanizing dynamic induced by this axis of infrastructures.

The research deals then with contrasts, frequently seen as unsolvable problems or conflicting conditions, that can so often be described as those that make up the Theatre of Memory, resistant to change – though it is change that guarantees their modernization and continuance. Here, the studio's research aims to reconcile contrasts by generating alternatives that show the potential of creative commitment.

A studio such as this one for Beijing serves to discover strategies that prove the opportunities of this location when the right techniques of urbanistic and environmental analysis are employed (as most of the students involved have understood), giving rise to a creative process that works simultaneously at various spatial scales and leads to thought-provoking strategies and projects that are full of innovative nuances. It is in this way that a studio of this kind can serve as a Theatre of Prophecy that offers platforms for innovative thought about proposals, in contrast with positions that are anchored merely in "Memory".

The exercise involved the participation of students from different departments within the GSD and a multidisciplinary team of teachers, ensuring broad-based monitoring of the students' work and a richer body of references and reviews. The participation of Professor Kongjian Yu from Beijing University and Professor Peter Rowe, specialist in urban development in Asia, guaranteed the students a systematic examination of the alternatives they produced. Additional instructors with experience in Asian contexts – Mark Mulligan (Department of Architecture) and Stephen Ervin (Department of Landscape Architecture) – provided continuous support for the development of the studio. The density of the work presented here is not simply a reflection of the enormous effort invested in the studio but also of the valuable experience it represented for participants.

The propositive simulation of studios, even at the territorial scale presented here, serves to validate the capacity of the "project" as a suitable instru-

本次课程探讨台湖镇的城市化和发展，目前正在形成的京津轴线是环首都城市体系中最重要的轴线之一，北京沿京津轴线的扩张正改变着台湖乡村地区。

课程重点研究北京动态城市系统在快速发展中产生的矛盾。一方面是环城公路与北京市城市总体规划中提出的第二道绿化隔离带模式的对照，另一方面是京津城际与京津高速辐射轴带来的城市化驱动力（目前在建高铁和高速公路使"带状模型"失衡）。另一个问题是在面对基础设施辐射轴带来的经济发展和城市化动力的同时，保护农业和生态价值高的区域十分困难。

通常被认为是无法解决的问题或博弈的现状，也可以描述为所组成的场所记忆，使人们拒绝去改变，即使这些改变能保证现代化和持续发展，我们将它们进行对比研究。在这里，课程的目标是通过产生具有创新潜力的多解方案去调解这些矛盾。

从运用城市和环境分析等技术寻求论证场地发展机会的策略，上升到一个富有创造性的过程，即同时研究不同的空间尺度，能带来发人深省的策略和充满创新思想的方案。应用这种方法能让课程作为产生创新思想的平台，而不是被过去的思想困住。

本次课程有来自哈佛大学设计学院不同科系的学生和多学科的师资队伍，确保能够广泛地指导学生工作和提供丰富的参考资料和意见。北京大学俞孔坚教授和亚洲城市发展研究专家彼得·罗教授的参与，保证了学生提出的各种方案得到系统的检验。其他导师，如具有亚洲研究背景的来自建筑学系的马克·马利根和来自景观设计学系的斯蒂芬·欧文教授，为课程运行提供持续的帮助。这里提到的工作不仅仅是为课程所投入的巨大努力，也代表参与人员获得的宝贵经验。

即使在这个场地尺度里，课程仍试着模拟验证以"方案"的形式是否能作为产生各种创新方法和改善环境的合适手段，这也是在北京研究课程中出现越来越多的一个重要的问题。

ment for generating innovative solutions and improving the environment, a vital question in the circumstances of exponential growth found in the Beijing Studio.

It is, then, a great pleasure to present the publication of this body of work with a view to launching a discussion about ways of designing the great metropolis that aspire beyond the conventional slogans currently besetting planning. This publication should be seen as a breath of fresh air and a proclamation of optimism for a more creative, responsible future.

Joan Busquets is the Martin Bucksbaum Professor of Urban Planning and Design at the Graduate School of Design.

我们很荣幸能出版这本书以期展开讨论建设大都市的方法，有志于超越目前困扰规划的传统口号。这本书的出版可视为一缕新鲜的空气和对更富创造性和责任感的未来的乐观宣言。

琼·布斯克茨是哈佛大学设计学院城市规划与设计系的马丁·布克鲍姆教席教授。

Studio
Introduction
课程简介

Studio Introduction
课程简介

Under the auspices of rising populations, improved material standards of living for many, and a significant enlargement of urban functions, cities in China have continued to expand into their hinterlands, often displacing agricultural production and encroaching upon natural and other conservation areas. Over the past thirty years, for instance, the population of Beijing has doubled to around 9.8 million in non-agricultural residents, 12.6 million in urban district dwellers, and probably a total of some 15.5 million counting floating (unregistered) and itinerant populations. During the same period, the built area of the city expanded close to six-fold, from roughly 310km² in 1980 to more than 1,800km² today. There has also been extensive expansion of Beijing's administrative districts – which at the moment number eighteen – in order to keep up with management issues. As development has pressed forward, significant challenges have emerged particularly towards the outer margins of the city, including environmental degradation (water shortage and water pollution, soil desertification, and poor air quality), diminution of arable land, and an erasure of associated cultural identity. In these general conditions and in many of the particulars, Beijing is not alone. Peri-urban development, with its inherent and unique mixture of urban and rural circumstances, is among the most problematic spatial conditions confronting China as it moves forward in its present round of modernization.

To investigate Chinese peri-urban development, its challenges and potentials, a group from Harvard University's Graduate School of Design undertook research in the spring of 2011 under the leadership of visiting professor Kongjian Yu and GSD professors Peter Rowe, Stephen Ervin, and Mark Mulligan, teaching assistant Har Ye Kan, and visiting scholar Hailong Liu. Twelve graduate students drawn from programs in architecture, landscape architecture, urban planning, and urban design served as primary investigators and designers. The research focused on alternative futures for the town of Taihu (Taihuzhen), situated in the Tongzhou district of southeastern Beijing, some twenty kilometers from the city center. Covering an area of approximately 36km², the town is comprised of 46 villages (of which 25 are located within our study area)., as well as spill-over urbanization from neighboring population centers, mainly on the western side of the town. Ad hoc arrangements of factories, typically as part of township-village enterprises (also not an uncommon feature of this kind of landscape in China), are often located next to the villages. As the lowest part of Beijing in elevation, the town of Taihu and its surrounding area also act as a drainage sink, the site of rich wetlands that are high in agricultural productivity, especially its lotus ponds. Much of this area was planned as a "green wedge" in Beijing's Comprehensive Plan, with road, rail, and other infrastructure rights-of-way serving today as green belts traversing through the town. In 2011, there are about 40,000 registered inhabitants in the town and an additional "floating population" of 80,000. Of this number, approximately half (or 60,000) reside within our study area, comprising 9% of the broader Tongzhou District population... though this seems likely to change in the near future. The newly opened Beijing-Tianjin High-Speed Railway traverses the site with a planned station stop (Yizhuang Station) at the center of the town, intersecting with the recently opened Yizhuang Line of the metropolitan transit system, departing from the central eastern

在人口不断增长，物质生活条件改善，城市功能显著扩大的情况下，中国城市向其周边不断扩张，经常侵占农业生产和自然及其他储备的地区。在过去的30年，北京市人口翻了一番，非农业人口达到980万、城市居民1250万、流动人口约1550万。与此同时，北京市建成区范围从1980年的310平方公里到今天超过1800平方公里，面积扩大了6倍。为了跟上管理问题，北京行政区也在外延扩张。随着发展的不断推进，给北京边缘地区带来许多巨大的挑战，包括环境恶化（水资源短缺和水污染，水土流失以及空气质量恶化）、耕地减少和文化认同感缺失，这种情况并不只发生在北京，城郊是城市与乡村环境固有和独特的混合体，其发展面临中国当前一轮的现代化中最严峻的空间条件。

为了探讨中国城郊发展、挑战和潜力，客座教授俞孔坚和哈佛大学设计学院教授彼得·罗，斯蒂芬·欧文和马克·马利根，助教简夏仪和访问学者刘海龙，带领哈佛设计学院一支研究队伍于2011年春天接受研究任务。来自建筑、城市规划和城市设计系的12位研究生作为主要研究人员和设计师，研究关注台湖镇不同的未来。台湖镇位于北京东南部的通州区，距离市中心大约20公里，面积约36平方公里，由46个村庄组成（其中25个位于我们的场地），来自相邻人口密集区外溢的城市化人口主要集中于台湖镇西侧，工厂尤其是乡镇企业位置紧邻村庄。作为北京海拔最低处，台湖镇及其周边地区犹如排水槽，拥有丰富湿地，特别是荷塘具有高农业生产力。北京市城市总体规划中将台湖大部分地区划为"绿楔"的一部分，当前的道路、铁路、公路和其他道路基础设施作为绿化带穿越城镇。2011年，台湖镇约有4万居住人口和8万流动人口，将近一半（约6万人）在我们研究的场地内居住，占通州区人口的9%，然而在不久的将来人口数量很可能会改变。

新开通的京津城际横贯场地并于镇中心设有停站点（亦庄站），与最近开通的亦庄地铁线相连，地铁线计划从北京中心东部地区扩展延伸到台湖镇域。这些改进将大幅促进场地的城市化发展，给农业和生态保护带来更大的压力。台湖镇出台了各种方案吸引

CONGESTED STREET WITHIN TAIHU TOWNSHIP

台湖镇拥挤的街道

side of the city and scheduled to continue on through Taihuzhen. These improvements will substantially enhance the urban-oriented accessibility of the area, bringing further pressure to bear on its use for agricultural and conservation purposes. The town authority is entertaining various schemes to attract international investment, including the construction of a 4-km^2 "new community" or, alternatively, a tourist-oriented and recreation-oriented townscape. In short, a number of competing claims are being brought to bear on Taihuzhen, requiring consideration and possible resolution.

Our research objectives for the study were twofold: the first was to develop a broad urban district-level proposal, involving landscape, urban and architectural strategies within the broader Beijing Metropolitan Region (and considering the relatively close proximity to neighboring Tianjin to the southeast); the second was to formulate and develop site-specific proposals within the scope of the broader district level plans that could appropriately amplify and particularize strategic dimensions and/or contextual conditions of those plans. Student participants were encouraged to adopt particular points of view, aimed at resolving competing claims in Taihuzhen. Among the targets of opportunity were 1) development intended to improve the quality of life of local residents; 2) conservation of local cultural assets and productive agricultural lands; and 3) interpretation and implementation of the "green wedge" and "greenbelt" strategy envisioned the Beijing Municipal Commission's plans.

We began with study area analysis and representation of critical development and preservation issues, including abiotic and biotic aspects of the physical environment; socio-economic factors and proposed improvements; building and landscape typologies supporting inhabitation and productive use; and likely interests and values associated with relevant position-taking and decision-making. Within socio-economic features, this included approaches to value capture, for instance, or property-rights transfers. In the domain of habitation and use, preferred typologies and evolutionary patterns of settlement have been investigated. Scenarios for administratively and politically navigating competing claims were pursued. Following an eight-day (February 18-26, 2011) field trip to explore conditions on the ground in Taihu, the initial analytical phase was followed by group proposals for district-level strategies for the Taihu study area; this phase was then followed by site-specific proposals and projects, which both illustrated details of and served as proof-of-concept for the district-level proposals.

国际投资，包括建设 4 平方公里的新社区和以旅游休闲为主导的城镇景观。总之，台湖镇的发展涌现了许多问题，需要仔细考虑和寻求可能的解决方案。

我们研究的目标是双重的：一是在大北京都市区域下（并考虑东南部相对邻近的天津）提出一个包括景观、城市和建筑策略的大城市区级方案。二是在大区级规划下制定和提出场地特定的方案，可适当地放大或注重那些规划的策略层面或环境条件。鼓励学生采取特定的观点解决台湖镇的发展问题。目标是：（1）提高当地居民的生活质量；（2）保护当地文化资产和具有生产力的农地；（3）诠释和实施北京市城市总体规划中提出的"绿楔"和"绿带"策略。

我们从场地分析和陈述关键的发展和保护问题开始，包括非生物和生物方面的物理环境、社会经济因素和改进的建议、支持居住和生产使用的建筑和景观类型与定位和决策可能相关的利益和价值观。在社会经济特征层面研究包括价值获取的方法和财产权的转让，在居住和使用层面，研究居住区的合适类型和演化模式，寻求从行政和政治上引导问题的愿景。在接下来 8 天（2011 年 2 月 18-26 日）场地调研探查台湖的地面状况，分析的初始阶段以小组形式提出台湖研究范围的区级策略，接下来是具体场地的提议和方案，需要阐述方案细节和验证区级方案的概念。

COLLECTING AND CLEANING RUBBLE FROM A RECENTLY RAZED VILLAGE   从新拆迁的村庄收集和清理废砖

Beijing Regional Analysis
北京区域分析

Context: China's Bohai Region
地理背景：中国环渤海地区

Metropolitan Beijing is tied, both economically and ecologically, to the greater Bohai Region, which extends from the mountains to Beijing's northwest down to Bohai Sea to the southeast. In addition to Metropolitan Beijing, the Bohai Region also includes the major port cities of Tianjin and Dalian and many industrial cities lining the coast from the Shandong Peninsula to Liaoning Province. The Chinese central government has invested heavily in consolidating the Bohai Economic Rim as an urbanized region essential for the nation's economic advancement. The region occupies a critical geographic position at the northernmost edge of the North China Plain, between the cold temperate zone to the north, arid desert lands to the west, and the productive agricultural lands to the south. Water scarcity has long been a concern in the region – for centuries, aqueducts have diverted water to Beijing from the rainier south. But water scarcity will only become more dire if current rates of desertification, aquifer depletion, population growth, and urban and industrial expansion continue. The long-term sustainability of Beijing's continued growth will depend on decisions taken, at all scales of development, to preserve the health of the greater Bohai Region.

北京大都市圈的经济和生态都与大环渤海地区紧密联系，大环渤海地区是从北京西北山区到东南的渤海。除了北京，环渤海地区包括主要的港口城市天津和大连，以及沿着山东半岛和辽宁省海岸的众多工业城市。中国中央政府投入巨资来巩固环渤海经济区，它是关系国家经济发展重要的城市化区域。该地区位于华北平原边缘，北部是寒温带，西面有干旱的沙漠地带，南部有肥沃的农地，占据一个至关重要的地理位置。几个世纪以来，缺水是这个区域一直被关注的问题，北京利用管道从多雨的南方输送水资源。如果土地沙漠化、蓄水层枯竭、人口增长及城市和工业扩张速率持续，缺水问题只会变得更严重，北京的长期稳定持续发展将依靠保护大环渤海地区健康发展的决策。

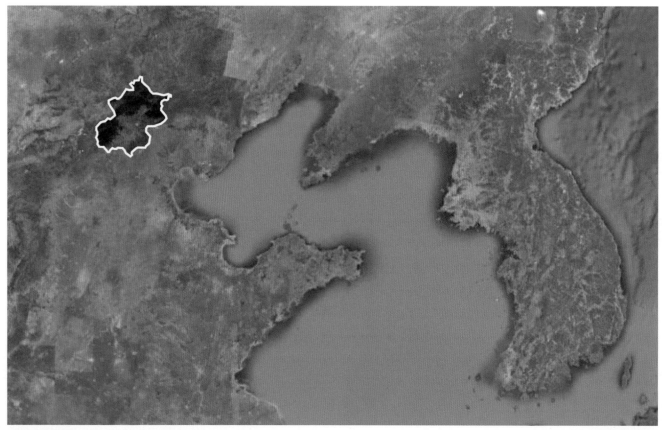

CONTEXT OF THE BEIJING REGION AND BOHAI SEA

北京与渤海的地理位置关系

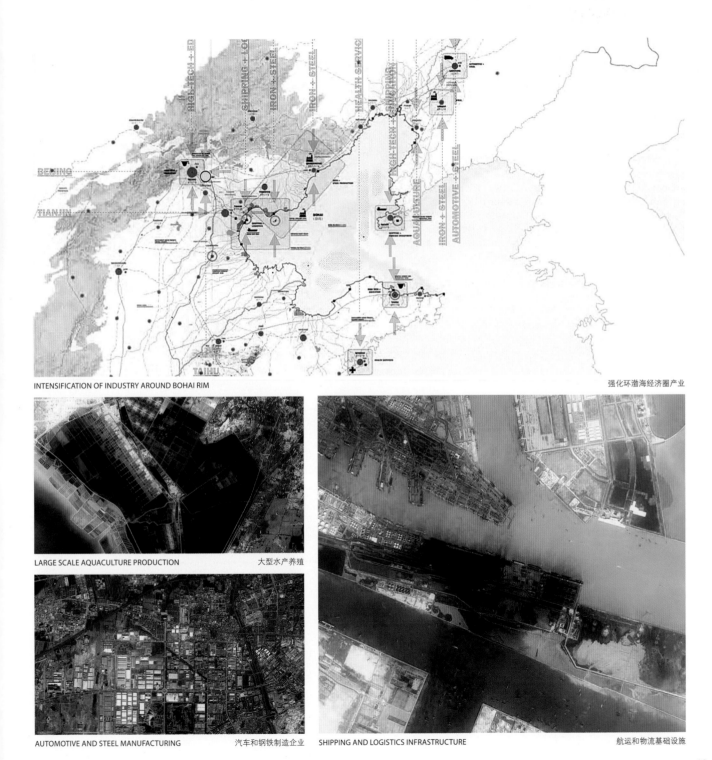

INTENSIFICATION OF INDUSTRY AROUND BOHAI RIM　　强化环渤海经济圈产业

LARGE SCALE AQUACULTURE PRODUCTION　　大型水产养殖

AUTOMOTIVE AND STEEL MANUFACTURING　　汽车和钢铁制造企业

SHIPPING AND LOGISTICS INFRASTRUCTURE　　航运和物流基础设施

Beijing's Growth and Peri-urban Issues
北京的城市扩张和郊区问题

Beijing's explosive 20th and 21st century expansion is revealed in the way that its original Ring Roads – the First (from the 1920s, no longer visible in the city fabric), Second (1980s), Third (1990s), and Fourth (2001) – which once defined concentric growth boundaries, are already fully incorporated into dense urban fabric. The recently completed Fifth and Sixth Ring Roads (approximately 10 km and 15-20 km from the center, respectively) now establish the perceptual boundaries of a metropolis six times larger than Beijing was thirty years ago. As annular urban development extends further from the city center, however, the strength of historic abstract geometric planning ideals begins to break down, and other factors, such as local topography and proximity to transportation infrastructure and other amenities, become more important in shaping development. With the growth of industrial cities along the Bohai Sea and the introduction of new highways and a high-speed rail link connecting Beijing to coastal Tianjin, a new model – the "linear metropolis" – begins to suggest itself in the intensity of development within Beijing's southeastern periphery where our study area of Taihuzhen lies. At a local level, peri-urban development typically entails demolishing extended swaths of rural villages to create a tabula rasa for much higher-density residential and commercial development. In well situated areas like Tongzhou District, the potential rewards for real estate speculation are particularly high; here uncontrolled rapid development has resulted in the loss of community fabric and serious environmental degradation.

环城公路的建设表明北京从 20 世纪到 21 世纪爆炸性的膨胀：第一条（1920 年代，在城市结构中已不可见）、第二条（1980 年代）、第三条（1990 年代）、第四条（2001 年）——曾经定义为同心圆扩张的边界，已经完全融入密集的城市肌理。最近建成的五环和六环路（大约各自距离市中心 10 公里和 15～20 公里）形成可感知的大都市边界，围合的面积是 30 多年前的北京的 6 倍。城市逐年扩张，传统抽象的几何规划理想开始瓦解，其他的因素如当地地形和邻近交通运输基础设施及其他设施，对其发展更为重要。随着渤海沿岸工业城市的发展，以及连接北京和天津滨海区的高速公路和高铁的建成，北京东南外缘即我们的研究区域台湖镇所在，在其高速发展中形成一种新的"线性城市"模式。在地方层面，典型城郊的发展需要拆除大片农村用来建造高密度住宅和商业开发，如通州区不动产、投机买卖潜在的回报甚高。这里不受控制的快速发展会带来社区结构缺失和环境恶化。

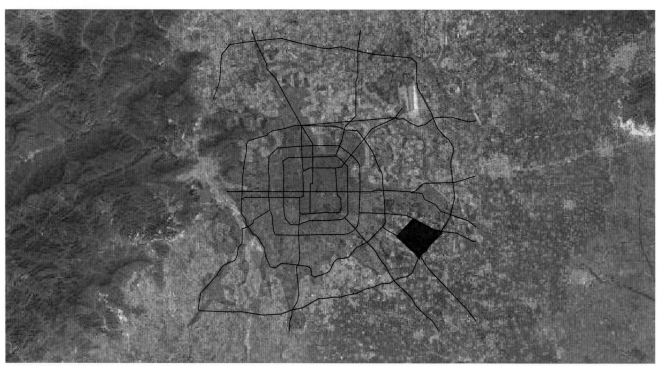

TAIHU STUDY AREA WITHIN BEIJING'S SPRAWLING CONCENTRIC GROWTH PATTERN

台湖研究区域位于北京同心扩张区

PROXIMITY OF BEIJING AND TIANJIN URBAN EXTENTS
与北京和天津的城市范围的距离

EXISTING VILLAGES RAZED IN FAVOR OF NEW CONSTRUCTION　现有村庄被夷为平地以便于新的建设

TYPICAL SPRAWL ENABLED BY THE EXTENSION OF HIGHWAYS INTO THE HINTERLAND　高速公路延伸到腹地导致典型的扩张

"Green Wedges" in the Beijing Metropolitan Plan
北京市城市总体规划中的"绿楔"

The idea of green belts – continuous expanses of protected, unbuilt landscape to contain urban sprawl and ensure proximity of city-dwellers to areas of scenic beauty and recreational potential – enjoys a long history in city planning of Western Europe and North America. Beijing's history of concentric outward growth might have suggested a natural fit for the green belt model when it was first proposed by Chinese planners in the 1950s; however, insufficient protections against development ultimately resulted in the First Green Belt's becoming fragmented and discontinuous. In 2003, parallel to the development of the Sixth Ring Road, a Second (outer) Green Belt Plan was approved; however, within a few years – in recognition of development pressures in the metropolis' essentially radial growth – city planners transformed the Green Belt concept into a regional "Green Wedge" strategy. As continuous ecological corridors connecting inner Beijing to its rural hinterlands, the green wedges envisioned by the city's Metropolitan 2004-2020 Plan seem well suited to current patterns of development. Yet unless well defined, sustainable provisions for protection and use are implemented, over time individual wedges inevitably face development pressure as their surroundings densify. A thorough study of potential overlapping or complementary open space uses – including agricultural production, recreational parks, alternative energy production, environmental remediation, and so on – together with public education programs, may enhance the long-term viability of the current Green Wedge strategy.

绿带的概念在西欧和北美城市规划中历史悠久，指的是用来保护连续区域，防止城市扩张和确保城市居民能享有美丽的景观和休闲游憩。北京同心圆外扩的历史与1950年代中国城市规划者首次提出的绿带模型契合，然而城市发展及保护不足最终导致第一道绿化隔离带变得支离破碎和不连续。2003年，与六环路平行的第二道绿化隔离带规划实施。经过数年的发展，城市规划者认识到城市环形辐射扩张带来的发展压力，提出将绿带变成区域性"绿楔"的策略。北京市城市总体规划（2004-2020年）提出绿楔概念，适应城市当前的发展模式，通过连续的生态廊道连接北京内部和偏远农村。除非实施保护和使用的可持续发展条例，否则随着时间推移单一的绿楔系统不可避免地将承受周边高密度发展所带来的压力。我们此次深入研究探讨了重叠或补充开放空间的使用方式，包括农业生产、休闲公园和可替代能源生产、环境恢复等，结合公共教育，可以提高当前绿楔策略的长久可行性。

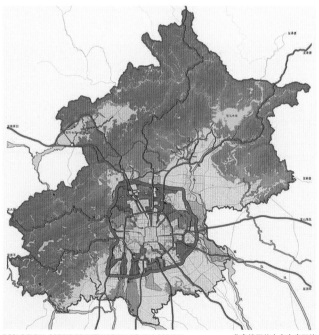

ECOLOGICAL CORRIDOR NETWORK IN BEIJING REGION　北京地区的生态走廊网络

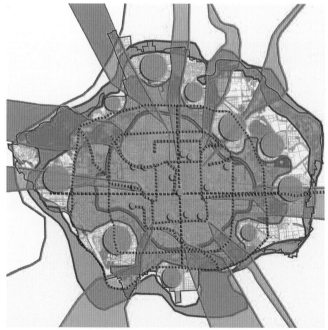

GREEN WEDGES AS PROPOSED IN BEIJING'S CITY PLAN　北京城市规划中提出的绿楔

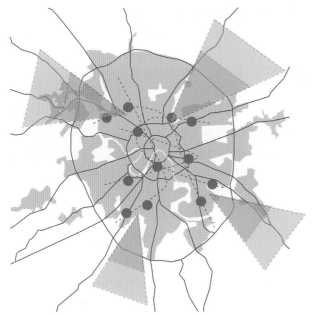

URBAN PLAN OF MOSCOW WITH GREEN WEDGES
莫斯科城市规划中的绿楔

TORONTO'S "GREATER GOLDEN HORSESHOE" REGIONAL GREENBELT
多伦多的"大金马蹄"区域绿地

UNPROGRAMMED OPEN SPACE AT BEIJING'S URBAN PERIPHERY
在北京城市边缘未规划的开放空间

Urban Adjacencies: Tongzhou and Yizhuang
城市邻接：通州和亦庄

As Beijing's urbanization approaches and surpasses the limits of the Sixth Ring Road, a number of "satellite city centers" have been proposed to help structure peri-urban growth. The two of these closest to Taihuzhen are Tongzhou New City to the northeast and Yizhuang New City (also known as the "Beijing Economic and Technological Development Area"), to the southwest. Envisioned as a new "world city" with a projected population of one million residents, Tongzhou is already highly urbanized and features a mix of residential, office, and commercial uses. Relatively close to Beijing's international airport and well served by public transportation, Tongzhou New City aims to compete with central Beijing in attracting international business; more recent condo development to the east is aimed at an upscale market. To the southwest, the planned urbanization of Yizhuang as a new center of economic and high-tech industrial development is still in early stages. Rural villages once occupying this area have already been demolished and residents relocated to new high-rise apartment buildings; the construction of Yizhuang's central business district has awaited the completion of the high-speed rail line linking Yizhuang Station to Beijing and Tianjin. Both of these new centers of urbanization rely on new transportation infrastructure to attract new businesses and residents; once well established, however, will these satellite communities sprawl out beyond current boundaries into the "green wedge" of Taihuzhen?

随着城市化进程和扩张超越六环路的限制，北京建设大量的"卫星城"推动城郊的发展。最靠近台湖镇的卫星城分别是位于其东北部的通州新城和西南部的亦庄新城。远景中的通州将成为百万人口的新世界城市，目前已经高度城市化，具有住宅、写字楼和商业混合使用的特征。通州新城地理位置相对靠近北京国际机场，具有便利的公共交通，其目标是能与北京中心区竞争吸引国际商务，近期东部地区高档住宅区的开发针对高端市场。在西南部，亦庄城市化区域的规划仍在初期阶段，这一区域将作为新的经济中心和高科技工业中心。原来的村庄已被拆除，居民迁移到高层公寓，一俟京津城际铁路连通亦庄，亦庄中心商务区将开始建设。这两个城市化新中心依赖新的交通基础设施来吸引新的企业和居民，然而，一旦设施成熟，这些卫星社区的扩张是否会超出目前边界，侵蚀台湖镇的"绿楔"？

PROPOSED PLANS FOR YIZHUANG AND TONGZHOU IN RELATION TO TAIHU STUDY AREA

提出的规划将亦庄和通州联系到台湖研究区

COMMERCIAL STREET FACADES WITH RESIDENTIAL TOWERS BEHIND 居民大楼正面对着商业街

CONGESTED STREETS 拥挤的街道

NEWLY CONSTRUCTED RESIDENTIAL TOWERS ARE TRANSFORMING YIZHUANG'S IDENTITY 新建的居民大楼正改变亦庄的地貌

Taihu Site Analysis
台湖场地分析

Land Use in Taihu
台湖的土地利用

Taihuzhen (Taihu Township) lies within Tongzhou District at the southeastern edge of urban Beijing and comprises roughly 36 square kilometers of land. Despite its location amid some of Beijing's most intensive peri-urban sprawl, Taihu has to date resisted the kind of "instant city" development that has characterized its neighbors Tongzhou and Yizhuang to the northeast and southwest. A majority of its land area lies within the boundaries of a Green Wedge defined by the Metropolitan 2020 Plan. Our study focuses on this central portion of Taihuzhen, an area chosen to coincide with four clear infrastructural boundary lines: the Jingjin Expressway and high-speed rail line to the southwest, the Tongma Road to the northwest, the Jingha Expressway to the northeast, and the Liangshui River to the southeast. Within this area, land use and demographics over the past two decades have not been static. Village boundaries have shifted or vanished, paved surfaces have increased, and light industry has sprouted up on previously agricultural lands. Irrigated agriculture reached a peak here in the early 1990s and has been slowly declining since; a shift to higher-value crops and new business models, however, continue to make it profitable. Competing claims on Taihu's future – from the town government, from village collectives, from established local businesses and new ones hoping to move in – will demand a greater effort at comprehensive land use planning to help Taihu achieve economic prosperity while fulfilling its ecological role as a Green Wedge within Metropolitan Beijing.

台湖镇坐落于北京城区东南边缘的通州区，面积约36平方公里。尽管处于北京最密集的城郊扩张地区，台湖迄今抵制了"即时城市"模式——类似其东北部通州和西南部亦庄的发展。大部分土地位于北京市城市总体规划中确定的绿楔内。我们的研究聚焦在由四条清晰的基础设施边界线围合而成的台湖镇中心区域，这四条边界分别是西南的京津高速和京津城际、西北的通马路、东北的京哈高速以及东南的凉水河。过去二十年，区域内土地利用和人口统计数据一直在变化。村界变化或者消失，硬质地面增加，轻工业厂如雨后春笋般出现在以前的农田上。灌溉农业在20世纪90年代初达到高峰，之后一直缓慢下降，转型为高经济价值作物或新商业模式，持续盈利。镇政府、村集体、当地企业和希望进入该地区的人都在营造台湖的未来，这需要努力制定土地利用总体规划，促进台湖在经济繁荣的同时扮演好作为大北京绿楔一部分的生态角色。

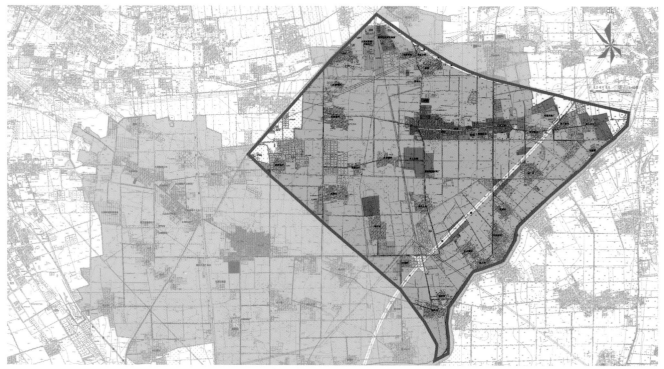

TAIHU STUDY AREA - LAND USE PRE 2008

台湖研究区域 -2008年前的土地利用

CURRENT LAND USE AND THE EMERGENCE OF INDUSTRY　　土地利用和工业现状

RECENTLY CONSTRUCTED CENTRAL TAIHU　　台湖中心区现状

TYPICAL MONO-FUNCTIONAL INDUSTRIAL CLUSTER　　典型单一功能组团

2009 PROPOSAL FOR TOWN CONSOLIDATION　　2009 年台湖镇整合方案

EXISTING VILLAGE UNDERGOING IMPROVEMENTS　　正在改造中的村庄

Hydrology
水文

Water – its quantity, quality, location, distribution, and use – is near the top of the list of urbanization issues around the globe, and no less so in China. In northern China, aquifer depletion is proceeding at unsustainable rates. Beijing receives an average of 620mm of precipitation annually, though due to significant fluctuations year to year, both droughts and seasonal flooding are common. Lying at the lowest point within Beijing Municipality (20 to 23 meters above sea level), the township of Taihu has relatively abundant surface water resources; its hydrological network includes the Xiaotaihou and Liangshui Rivers, canals, ditches, ponds and channels. A historic network of irrigation canals, dikes, and dams supports a range of agriculture and aquaculture; lotus root ponds and fish farms are common. Surface and subsurface water pollution is a major issue throughout the region: sewage from upstream settlements and sediments and chemical pollution from agriculture and industry are concentrated in this area. New developments need to contain and control surface water and floods while minimizing the introduction of impervious surfaces; mechanical and biological systems must be designed to counter surface and subsurface water pollution; and new policies and controls should be enacted regionally to reduce point-source and non-point-source noxious pollutants. The historic sweet spring water and clear-flowing rivers of the Beijing region must be restored and shepherded for future generations.

在世界各地城市化的议题中，水量、水质及水的位置、分配和利用等问题，都几乎位居榜首，在中国也是如此。中国北方水土流失严重。北京的年均降水量为620毫米，由于年波动较大，常见干旱和季节性洪水。台湖位于北京市内的最低点（海拔20～23米），地表水资源相对丰富，其水文网络包括萧太后河、凉水河、渠、沟、堰和水道。历史悠久的灌溉水渠、堤、坝网络支撑着一系列农业和水产业，荷塘和养鱼场十分普遍。地表和地下水污染是整个区域的主要问题：上游居住区排出的生活污水，工农业排放的沉积物和工业化学污水都集中在这一区域。新的发展需要控制地表水和洪水，同时尽量减少不透水地面；工程和生物系统的设计必须能应对地表水和地下水污染；新政策和控制措施应通过制定地区条例减少区域点源和非点源有毒污染物。必须恢复北京地区历史上甘甜的泉水和清澈的河流，留予子孙后代。

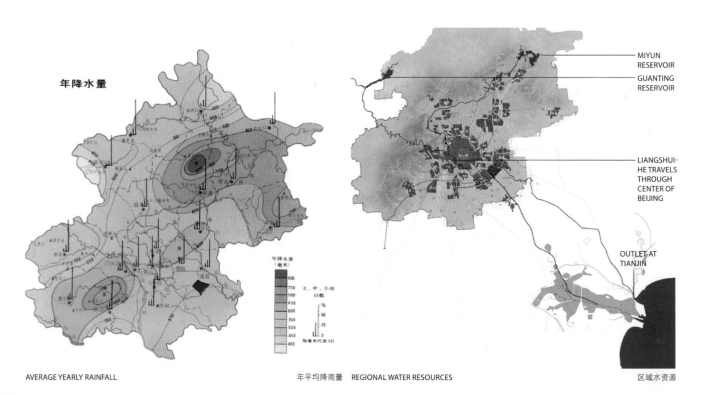

AVERAGE YEARLY RAINFALL　　　年平均降雨量　REGIONAL WATER RESOURCES　　　区域水资源

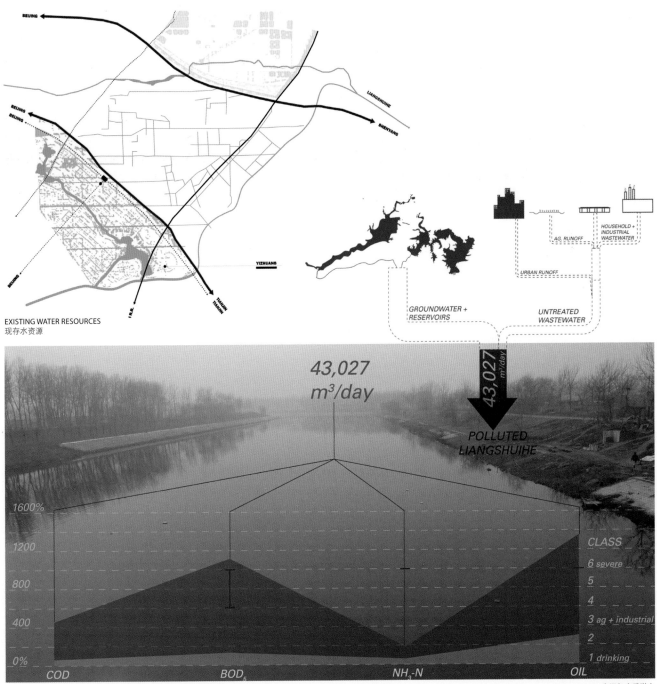

EXISTING WATER RESOURCES
现存水资源

SOURCES OF WATER AS RELATED TO WATER QUALITY POTENTIAL

水源与水质潜力

Agriculture
农业

China possesses only 7% of the world's arable land (mainly in the east) and fresh water (mainly in the south) while needing to feed 22% of the world's population. This circumstance has made food security a critical consideration in regional planning. The region surrounding Beijing lies at the intersection of the North China Plain's two major grain-producing agricultural belts – for winter and spring wheat – and maintaining the proximity of food production to major cities continues to be of primary importance. In recent decades, Taihu and other agricultural towns close to the capital have witnessed a shift in agricultural production away from wheat towards more valuable, labor-intensive crops such as fruits and vegetables (in particular, lotus root) and aquaculture (fish farming) – all of which involve water-intensive production methods familiar to farm laborers from regions further south. Taihu's low-lying topographic position has historically enabled it to channel water from the surrounding region into a network of irrigation canals to support intensive production; yet water pollution here and elsewhere in peri-urban Beijing is largely unregulated and represents an increasing concern for health risks in food production. Recently, modern agricultural enterprises such as Jin Fu Yi Nong (Tomato United Nations), with its dual focus on advanced production methods and gastro-tourism, have added new vitality to the local economy; the increased profitability of such enterprises may hold the key to financing infrastructural improvements that will sustain healthier, more productive organic farming in Taihu's future.

中国只拥有世界7%的耕地（主要在东部）和6%的淡水（主要在南部），而需要养活占世界22%的人口。粮食安全成为区域规划中的重要考虑因素。北京周边地区位于华北平原的两大粮食产区（冬小麦和春小麦）的交汇处，确保粮食产量满足主要城市需求是重中之重。近十年来，台湖以及其他靠近首都的农业城镇见证了农业生产的转型，从小麦转变为更有价值、劳动密集型的农作物，如水果、蔬菜（特别是莲藕）及水产养殖（养鱼），所有这些都涉及耗水型的生产方式，常见于南方地区。历史上，台湖利用低洼的地势从周边地区向灌溉沟渠网络引水，支持集约化生产。但这里和北京城郊其他地方的水污染不受监管，食品安全受到越来越多的关注。近来，现代农业企业，如金福艺农（番茄联合国）致力于先进的生产方法和美食旅游，为当地经济增添了新的活力。未来，越来越多此类盈利企业也许是支持台湖更健康、更高产的有机农业的关键。

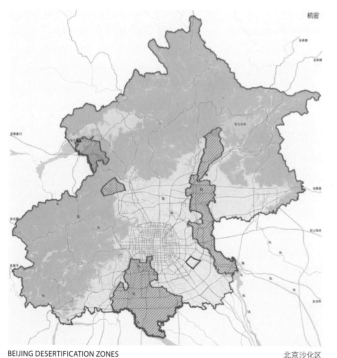

BEIJING DESERTIFICATION ZONES　　北京沙化区

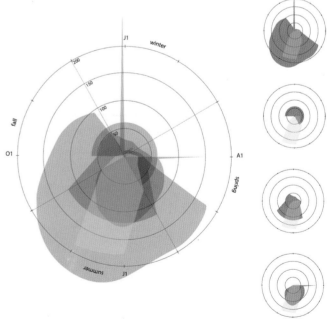

WATER REQUIREMENT / AVAILABILITY STUDY　　水需求和可行性研究

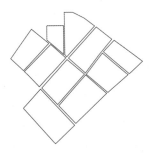 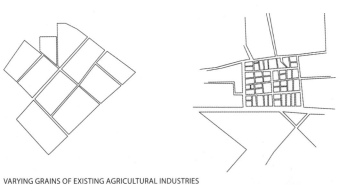 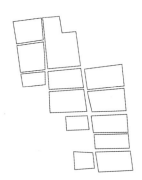 

VARYING GRAINS OF EXISTING AGRICULTURAL INDUSTRIES　　　　　　　　　　　　　　　　　　　　　　　　　现有农业的不同肌理

GREENHOUSES EXTEND THE GROWING SEASON　　　温室延长种植季节　　EXPANSIVE WHEAT AND RICE FIELDS　　　广阔的麦田和稻田

AUTOMATED MONITORING TO INCREASE YIELDS　　　自动化监测使产量增加　　LOTUS FIELD　　　荷塘

Nature Conservation
自然保育

While much of the planning around Beijing's peri-urban expansion is focused on development, infrastructure, basic needs, and economic growth, the importance of nature conservation is also recognized, both in the very "Green Belt" and "Green Wedge" planning concepts and in China's National Biodiversity Strategy and Action Plans. The multiple roles played by natural environments in providing wildlife habitat, human recreation and rejuvenation, and air and water cleansing are well known. Landscape ecological principles underscore the value of biotic diversity to landscape vitality and resilience and emphasize the need for three basic kinds of patterns in a healthy, functioning landscape: 1) large contiguous "patches" such as forests, fields, and wetlands; 2) linear "corridors", often along waterways or transportation paths; and 3) "stepping stones" or "islands" – smaller isolated natural patches that integrate with the built fabric, providing all the benefits of nature in the midst of development. While diverse mammalian wildlife is not evident in most of the Taihu study area, a variety of birds and other organisms are present, though endangered by expanding urbanism, environmental degradation, and habitat loss. Embedding, restoring, and protecting natural resources is an essential part of a well balanced, sustainable development strategy.

虽然北京大多数城郊扩张的规划关注发展、基础设施、基本需求和经济增长，但无论是在"绿带"和"绿楔"的规划概念还是中国生物多样性保护战略与行动计划中，自然保护的重要性都得到了认可。众所周知，自然环境在提供野生动物栖息地、人类游憩和康复以及空气和水净化中扮演了多种角色。景观生态学原理强调生物多样性对于景观生命力和恢复力的价值，强调健康的功能性景观需要三种基本格局：（1）大而连续的"斑块"，例如森林、田地和湿地；（2）线性"廊道"，通常沿着水路或者交通线路；（3）"跳板石"或"岛屿"，较小的独立自然斑块与既有肌理保持一致，在发展中提供自然的所有益处。哺乳类野生动物在台湖大部分研究区域并不常见，各种鸟类和其他生物由于城市扩张、环境恶化、栖息地丧失的威胁而濒临灭绝。嵌入、恢复和保护自然资源是均衡可持续发展战略的重要部分。

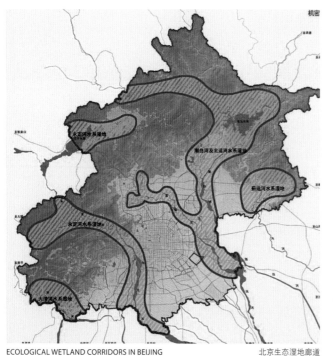

ECOLOGICAL WETLAND CORRIDORS IN BEIJING　北京生态湿地廊道

EXISTING WETLANDS TO FORM POTENTIAL CORRIDORS　台湖镇形成潜在廊道的现存湿地

NATIVE SPECIES IN NEED OF RE-ESTABLISHMENT 亟待恢复的本地物种

POCKETS OF DIVERSE VEGETATION CREATE HABITAT 多样植被斑块创造栖息地

Economy and Demographics
经济与人口

Due to availability of inexpensive rental units and convenient transportation network, Tongzhou District (of which Taihuzhen is a part) has one of metropolitan Beijing's highest concentrations of unregistered, itinerant workers – the so-called "floating population". According to data from 2010, Taihu has 19,000 households and a total population of 120,000 – of which only 40,000 are officially registered, while the remaining 80,000 are "floating". In 2009, average per capita rural income was 12,000 Yuan ($1,760), representing a 12% increase over the previous year. The township's economy, traditionally based on farming, continues to diversify as new industrial enterprises have been established in close proximity to agricultural fields. Ninety-two new enterprises were established in Taihu in 2009, boosting total industrial production to 9.55 billion Yuan (approximately $1.4 billion), and the town received tax revenues of 1.29 billion Yuan ($190 million). Agricultural production has also improved significantly in recent years with the introduction of subsidies, advanced farming techniques, specialized cultivation, and modern business operation models such as Jin Fu Yi Ning (Tomato United Nations). Yet the town's economic transformation has not benefited all residents equally; concerns abound for the rural poor and elderly, whose access to the evolving job market is limited by the availability of training programs.

由于提供了廉价的出租单元和便捷的交通网络，通州区（包括台湖镇）是大北京流动人口最集中的区县之一。2010 年，台湖常住户数 1.9 万户，总人口 12 万，其中仅 4 万为常住人口，其余 8 万为流动人口。2009 年，农民人均劳动所得达到 12000 元（1760 美元），同比增长 12%。乡镇经济过去以农业为基础，随着新的工业企业不断在农业用地附近落地，经济持续多样化发展。2009 年台湖引进企业 92 家，为全镇经济增长提供强劲动力，总工业产值达 95.5 亿元（约 14 亿美元），台湖税收总额为 12.9 亿元（1.9 亿美元）。近年来，通过引入政府补贴、先进耕作技术、专业化种植养殖及现代商业经营模式如金福艺农（番茄联合国），农业产量显著提升。然而，镇域经济转型并未平等惠及所有居民；应关注农村贫困人口和老年人，其进入就业市场的发展受限于职业技能培训的获得。

A SHORT SIGHTED REALITY　　　　　　　　　　　　　　　　　　　　短视的现实

A SHORT SIGHTED REALITY　　　　　　　　　　　　　　　　　　　　短视的现实

VILLAGE SCALED COMMERCE 村级商业

CITIZENS CONGREGATING IN THE PUBLIC REALM 市民聚集在公共区域

INFORMAL COMMERCE IN OPEN SPACE 开放空间中的非正式商业

AGRICULTURAL DRIVEN ECONOMIES 农业驱动型经济

IMPROVED AGRICULTURAL PRODUCTION 改良的农业生产

Transportation System
交通系统

Beijing's robust and fast-growing public transportation system continues to enable the city's rapid outward growth, and peri-urban Taihu appears particularly well connected to Beijing's historical core and Central Business District. In addition to its access to the Beijing-Tianjin Intercity Express (high-speed rail line) via the soon-to-be-opened Yizhuang Station, the study area will also be served by the L2 (Yizhuang) light rail line, connecting to south-central Beijing, and – further in the future – the planned S6 suburban railway line, connecting the satellite cities of Yizhuang, Tongzhou, and Shunyi. Our study area is bounded by two major expressways, the Jingjin (Beijing-Tianjin) Expressway to the southwest and the Jingha (Beijing-Harbin) Expressway to the northeast, while the Sixth Ring Road cuts across the southeastern portion of the study area, isolating a narrow band of agricultural villages along the Liangshui River from central Taihu. Despite the diagrammatic clarity of its major road network, Metropolitan Beijing continues to suffer from severe traffic congestion, and Taihu and its neighbors are not immune to its effects. Much work remains to be done in recalculating capacities of local road networks to enhance flow of long-distance routes, while taking care not to create hostile environments for pedestrian and bicycle traffic. From an ecological preservation point of view, Taihu's current situation of relative inaccessibility may be something of a virtue. For the future, a careful balance must be maintained to ensure some degree of access to the outside without encouraging increased through-traffic between Tongzhou and Yizhuang.

北京强劲发展的公共交通系统促进城市迅速扩张，城郊的台湖与北京内城和 CBD 得以良好连接。除即将开通的京津城际（高速铁路线）亦庄站外，连接北京南城的轻轨 L2（亦庄）线，规划中连接亦庄、通州、顺义等重点新城的市郊铁路 S6 线，都将服务于研究区域。我们的研究范围以两大高速公路为界，即西南的京津（北京—天津）高速及东北的京哈（北京—哈尔滨）高速，而六环路横穿研究区域的东南部，将凉水河边呈狭窄带状的若干村庄与台湖中心隔离开。尽管主要道路网络清晰，大北京却继续遭受着严峻的交通拥堵问题，台湖及其周边地区也未能幸免。需重新计算当地道路网络容量，提高长途路线客运量，同时注意避免影响行人和自行车交通。从生态保护的角度来看，目前台湖的位置相对偏远可能是件好事。未来必须保持平衡，在不鼓励增加通州、亦庄之间交通的情况下，确保台湖与外界的适度连接。

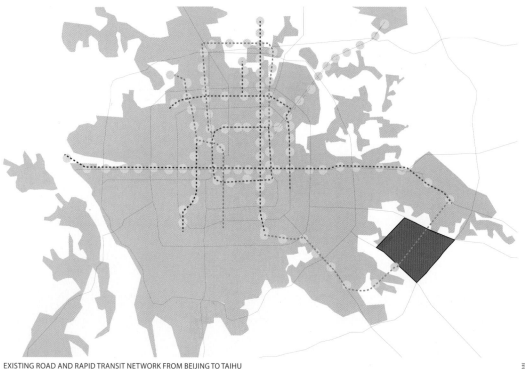

EXISTING ROAD AND RAPID TRANSIT NETWORK FROM BEIJING TO TAIHU

现有北京至台湖的道路及快速交通网络

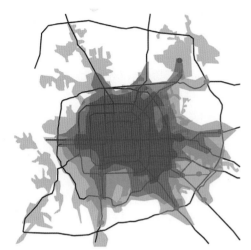

BEIJING CBD TO TAIHU ACCESSIBILITY
北京 CBD 至台湖的可达性

TAIHU TO BEIJING ACCESSIBILITY
台湖至北京的可达性

BUS SERVICE IS UTILIZED HEAVILY 过度拥挤的公交

THE SUBWAY STATION NEAR NEWLY CONSTRUCTED HIGH SPEED RAIL STATION 新近修建的高铁站附近的地铁站

EXISTING ROADWAYS HAVE BECOME CONGESTED 现有街道拥挤不堪

HIGH SPEED RAIL CONNECTING BEIJING AND TIANJIN 连接北京和天津的高速铁路

Settlement Patterns
居住模式

In 2011, Taihu offers its 120,000 residents two basic types of housing: the one-storey courtyard houses (siheyuan) of traditional villages and the modern high-rise apartment buildings that are steadily replacing them. For many, these two housing types simply represent the past and future of Chinese lifestyle. Many of Taihu's existing courtyard houses are not in optimal condition: lacking modern amenities, broken up into several apartments, and far more densely inhabited today than in the past. When aggregated into villages, this housing type does provide well scaled, pedestrian-friendly street space and a connection to the area's cultural history; however, villages whose dense street patterns are not easily integrated with Taihu's modern road network exhibit varying degrees of isolation. The new apartment towers offer a rational model compatible with increased population density as well as vehicular and sunlight access; when those same planning rules are applied to repetitive arrays of towers, however, the result is frequently an impoverished model of settlement in which residents are isolated from communal life and pedestrians lack connection to landscape and street life. Fresh ideas are needed for Taihu's planning to meet the needs of current and future residents – including for example: 1) creating density gradients across the land and concentrating new development around nodes of public transportation; 2) introducing more diverse housing types such as medium-rise (2-6 storey) courtyard-based housing; 3) preserving and renovating the traditional fabric of villages; and 4) providing a richer experience for pedestrians and bicyclists in the planning of streets, pathways, and access to nature.

在2011年，台湖12万居民有两种基本的住房类型：传统的四合院和现代高层公寓楼，后者正在取代前者。对很多人而言，这两种住房类型仅仅代表了中国过去与未来的生活方式。台湖许多现存的四合院现状堪忧：缺少现代化的设施，被分割为若干单元，比过去居住得更为拥挤。当集聚为村庄时，这种房屋类型的确提供了好的尺度感、行人友好的街道空间及与该地区文化历史的衔接；然而，村庄密集的街道格局难与台湖现代化的街道网络相融合，反映出不同程度的隔离。新建高层公寓提供了一种合理的现代居住模式，既与增长的人口密度相协调，又与车行和日照要求兼容。而当这些同样的规划准则应用于重复排列的高楼时，结果常常是乏味无力的居住模式，居民从公共生活中被隔离，行人缺少与景观及街道生活的联系。台湖的规划需要新的理念，以满足当前及未来居民的需求。例如：（1）创建土地密度梯度，集中围绕公共交通节点发展；（2）引进更多样化的房屋类型，如中等高度（2-6层）的庭院住房；（3）保留并改造村庄的传统肌理；（4）规划街道、小径通向自然，为行人及骑自行车的人提供更丰富的体验。

CONTRASTING DEVELOPMENT STYLES: RECENT RESIDENTIAL SET BEHIND TRADITIONAL CONSTRUCTION

风格对比：传统建筑后的新建住宅

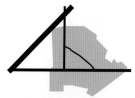 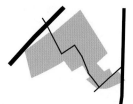

VILLAGE SITING - INFRASTRUCTURAL WEDGE INTERSECTION
村庄选址——楔形路口的基础设施

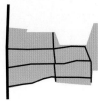 

VILLAGE SITING - INFRASTRUCTURAL ADJACENCY MAGNET
村庄选址——临近干道的基础设施

INTERNAL VILLAGE STRUCTURE   村庄内部结构

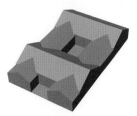  

TRADITIONAL HOUSING TYPOLOGIES   传统住宅类型

INTER-VILLAGE COLLECTORS   村间状况

INTER-DISTRICT ARTERIALS   跨区主干道

RURAL FARM ROADS   乡村田间道路

Future Prospects + Design Proposals
前景展望和设计方案

Essay: Web, Band, Network, and Field Operations:
Alternatives to the Conventions of Green Belts and Wedges in the Beijing Metropolitan Plan
by Peter G. Rowe

网、带、网络和场地运营：
北京市城市总体规划中绿带和绿楔的多解方案
彼得·罗

The urban environments of most if not all Chinese cities are spatially organized by state approved master plans. Generally these plans consist of spatial designations of acceptable uses, usually concentrated in central locations and in satellites out towards metropolitan peripheries; conspicuous radial and circumferential major roadways; and green spaces serving as nominally recreational and non-built spatial parentheses between areas of intensive urban activity. These green spaces, in turn, have often taken the form of "green belts" in the past, attempting to encircle and define broad areas of urbanization. Failures of these green belts to maintain a continuity and integrity of public open space, under pressures to simply build over them, gradually led in China to less constricting patterns of green space deployment, in the form of "green wedges" broadening out from central areas to the urban periphery. The celebrated plan of Suzhou, for instance, under the moniker of "One Body and Two Wings", gained its spatial clarity as much from the green wedges impinging on the joints of the "wings" with the "body", so to speak, as it did from particular patterns of urban development. Elsewhere, Piacentini's Wedge, named for the planner most responsible for its preservation, pushes into the very heart of Rome from its rural outskirts, comprised primarily of the Parco della Caffarella, an almost wild preserve dotted with ancient ruins. The coarse grain of these relatively vast and at times ill-defined vegetated areas can also be found in recent plans for other Chinese cities, ranging in scale and other specific characteristics from Kunming to Wuhan and even to Shanghai out towards its periphery. Perhaps nowhere, however, is the presence of green wedges a more emphatic element radiating out from the central city than in the current Metropolitan Plan for Beijing (2004-2020). This is probably due to the canonical annular symmetry of the city's spatial order over centuries of time-honored development and a continuing desire to preserve this centralized clarity in China's capital. Nevertheless, although very clearly visible in the plan, the more precise composition and inhabitation of these green wedges remains largely undefined, begging the question of their effectiveness in maintaining the desired spatial parentheses between locations of more or less intensive urbanization. Further, their functional and, indeed, aesthetic relevance as appropriate counterpoints to contemporary urban life also goes unanswered in the plan's current state, requiring re-visioning of the more broadly landscaped components of the metropolitan periphery such as those inherent to the present Taihu area. This task is further complicated by the exigencies of agricultural occupation and rural life, as well as, in the case of Taihu in particular, the mounting pressures of bordering infrastructure development, like the rapid inter-city rail, and neighboring building intensity from Tongzhou and Yizhuang. In response to the above conditions and contingencies, our research proposes a number of possible alternative futures for the Taihu area, organized within four identifiable and distinct group strategies.

大部分中国城市环境是依照国家批准的总体规划进行空间组织的。一般来说，这些规划包括可用空间设计，通常集中在城市中心和大都市边缘的卫星城；有明显的放射状和环状主干道；服务于城市密集活动区域之间，名义上的游憩和禁建绿色空间。这些绿地空间在过去常以"绿带"的形式出现，试图包围和定义广阔的城市化地区。这些绿带旨在保持连续、完整的公共开放空间，但迫于建设的压力未能成功，逐渐使得本应以"绿楔"的形式从中心地区向城市边缘扩大的绿地格局不完整。例如著名的苏州城市总体规划，在"一体两翼"思想的指导下，利用"体"和"翼"交界处的绿楔尽量增加空间通透感，这便是城市发展的特定模式。另外，以负责其保护的规划师命名的匹亚辛迪尼绿楔，从乡村郊区一直深入罗马的核心地带，包括一个古遗址星罗棋布、几乎是荒野的保护区——帕克卡法瑞拉的主要部分。这种相对广阔、未开发荒地的粗放扩张也出现在最近中国其他城市的规划中，比如昆明、武汉甚至上海都向周围扩张。但是，也许目前没有比北京市城市总体规划（2004-2020年）中心城市的辐射绿楔更有力的元素。这可能归因于几个世纪公认的环状对称的城市空间发展及中国首都保持中央集权的明确愿望。虽然规划清楚直观，但并未明确这些绿楔更具体的组织和定位，如何维持密集城区之间理想绿色空间的有效性仍是问题。此外，与当代城市生活相适应的功能与美学的关联在目前的规划中悬而未决，需要重新审视大都市边缘更广阔景观的组成，如目前台湖地区。此任务在急需农业职业和农村生活的情况下更复杂，特别是在台湖，承受着边界基础设施发展的强大压力，比如城际快速铁路和从通州到亦庄的建筑密度。针对上述情形和不可预见的状况，我们的研究为台湖地区提出了一系列多解方案，分别为四种明确的、不同的策略组合。

第一个方案来自赖德·皮尔斯、马克·厄普顿和特洛伊·沃恩，充分利用"网络"：不同种类网络的安排各有差异，这些错综复杂

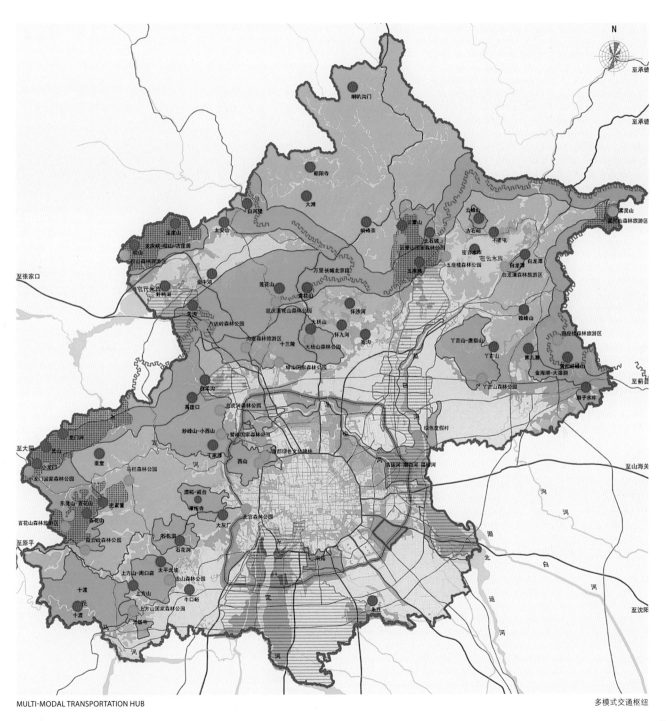

MULTI-MODAL TRANSPORTATION HUB 多模式交通枢纽

The first proposal, by Ryder Pearce, Mark Upton, and Troy Vaughn, makes very strong use of "network" arrangements, differing in degree as well as kind, from webs through interconnected infrastructural operations and complexities. Uppermost within the proposal's concerns is the water network and the theme of water security within an environment of scarcity. It is no secret, for instance, that Beijing is running ahead of its indigenous water supply sources and is located in an area of China that has habitually suffered from water scarcity. Throughout, the media and functions involved in the overall water network vary, but the coherence, connectivity, and key interrelationships remain intact. Across the board, the proposal is a very clear illustration of how water capture, cleansing, recycling and re-use can be significantly amplified in a water-scarce environment. The illustration also brings with it the potential added benefits of improved environmental quality and diversity. Similarly, the approach to built form is systematic and process-oriented, as well as being integrated into the overall networking logic and its intrinsic aesthetic outcomes.

The schemes of the second group, consisting of Yenlin Cheng, Hana Disch, and Aditya Sawant, offer clear alternatives to the green wedge in the form of a more diffuse green web of open space, with clearly defined programs of both use and spatial appreciation throughout its various strands and also serving as helpful armatures around which both agricultural and non-agricultural functions can be renewed and revitalized. A key feature throughout these proposals is making multiple use out of each strand of the web, so as to increase the number and variety of constituent claims that might be placed on them by both area inhabitants and users. The object is not only to increase levels of amenity but to also make the web more robustly defensible against unwanted incursions from other more aggressive and less publicly-spirited users. From a sub-regional perspective more or less confined to the study area, it is also important to deploy the grain and weave, as it were, of the web in a manner that clearly integrates readily and substantially into neighboring areas, moving both within and around metropolitan activities and uses.

The third scheme, by Darin Mano, Hyung Seung Min, and Carrie Nielson, arrives at a broad spatial strategy of "banding", orchestrated around the twin considerations of contingency and balance, while also resulting in a succession of more or less parallel spatial bands of dominant activity that are perpendicular to the green wedge originally envisioned in the Beijing Plan. Contingency comes by way of designating areas to be primarily urban where the likely pent-up demand to be urban is high and, conversely, designating areas for recreational and ecological conservation where these urban pressures are less and where the opportunities for the latter functions are high. Balance comes by way of being open to settlement and re-settlement, but in a manner that maintains village structures intact as a conspicuous element in the overall landscape. Assuming a scheme of highly intensive urban development essentially associated with the scheduled high-speed rail station, the proposal also makes use of a web-like repertoire of "open" or "green" spaces, ranging from parks of various sizes, to boulevards and canal landscapes.

的网由相互连接的基础设施形成。方案关注水系网络和环境安全下的水安全主题。例如，北京位于中国常年缺水的地区，当地供水缺乏。水系网络各部分的媒介和功能不同，但连贯性、联通性及主要内在关系保持不变。总体而言，该方案清晰阐明了水是如何收集、净化、回收和再利用的，这在缺水环境下很有意义。这些实例也告诉我们改善环境质量与多样性可能带来的收益。同样地，建成形式的方法是系统的、以过程为导向的，整合整个网络的逻辑及内在审美的结果亦是如此。

第二个方案是由陈燕霖、哈娜·迪施与阿迪亚·萨瓦特明确提出的一个绿楔的多解方案，形式为充满绿色的空间网络，通过轴线明确界定使用过程和空间识别度，也作为更新、振兴农业和非农业功能的骨架。方案的主要特点是多ామ化利用网络中的每条链，增加当地居民和住户可能提出的各种需求。目标不仅在于提高美化水平，也使网络更有活力，避免有攻击性和缺少公共精神的使用者的破坏。以次区域的视角来看，重构原有肌理也很重要，因为它很容易融入邻近地区，推动大都市内部及周围的活动与使用。

第三个方案由达林·马诺、闵玄胜和和卡莉·尼尔森提出，协调可能性和平衡性两个因素，规划"带状"空间策略，同时形成一系列几乎平行的主导活动空间，垂直于北京规划中的绿楔。若区域城市化压力较大，更可能成为设计中的城市地区；反之，若区域城市化压力较小且发展为游憩和生态保护区的可能性较高，则更可能成为设计中的游憩和生态地区。新建及重建居住区产生的平衡，作为整体景观中的一种突出元素，维持现有村庄结构不变。若此区域因为有高铁站而成了高密度的城区，方案也充分利用了网状的开放空间或绿色空间，包括各种尺度的公园、林荫大道及运河景观。

第四个方案由卡梅伦·巴拉戴、法蒂·玛苏德和安德烈亚斯·威格拉基提出，强调粮食安全和土地利用强度、密度对中国的重要性，通过叠加公共开放空间和保护区网络，构建密集高层建筑的城市发展，在农业生产中形成了多产的"田地"。在过去的十多年里，中国尤其是沿海富庶地区的高度城市化，超过了国家自给自足粮

The fourth proposal, by Cameron Barradale, Fadi Masoud, and Andreas Viglakis, emphasizes the importance to China of food security and the usefulness of density and intensity of land use in producing a rich "field" of agricultural production – overlain by a network of public open space uses and conservation areas and framed by relatively intensive, high-rise urban development. The push to higher levels of urbanization, especially within China's richer coastal regions, exceeded the nation's capacity for self-contained food production over a decade ago, and food security has remained a strategic concern ever since. Recognizing the importance, and perhaps primacy, of agricultural production within the emerging Bohai conurbation, this proposal reframes the Taihu site from a broad perspective of regional development well beyond the Beijing Metropolitan Plan before being imbuing it with more localized features. Rather than the essentially linear elements of "bands", "webs", and "networks", a modular-like spatiality is conjectured that could be aggregated and integrated as a completed 'field' of varying use and terrain. Further, its deployment remains open to locations potentially subject to more or less intensive levels of constructed, agricultural and conservation activities, as well as profiting in spatial presence and potential political congruence with strong yet diverse development interests from the sheer density, intensity, and range of proposed local activities.

Finally, through design exploration, these kinds of investigations raise questions about the sustained validity of the time-honored, but one suspects outmoded, planning concepts of "green belts" and "green wedges" in China's urban master plans. By delving further into these commonly accepted, although still ill-defined, practices of spatial provision, it becomes clear that more contemporary, locally nuanced, regionally-acceptable and complex conceptions of landscape development should be introduced, refined and developed. Moreover, just as urbanization comes to the forefront in China, so might the interface, particularly at its periphery, with non-urban uses and landscape conditions. In so doing, it may also be possible for China's laudable, long-standing, largely pre-modern and non-opposing conceptions of city – cheng – and countryside – xiang – to be returned harmoniously to contemporary life.

Peter G. Rowe is the Raymond Garbe Professor of Architecture and Urban Design at the Graduate School of Design, and also a Harvard University Distinguished Service Professor.

食的能力，粮食安全仍然是战略焦点之一。新兴环渤海地区农业生产的首要性得到认可后，该方案以超出北京市城市总体规划的广阔视野，重构了台湖。不同于线性的"带"、"网"、"网络"元素，设想类模块的空间，聚合成有不同用途和地形的完整场地。此外，该场地的布局对潜在的不同程度建设、农业和自然保护地区开放，也有利于空间展示及与强劲、多样发展利益的潜在一致性，这些利益来自一系列高密度、强度的当地活动。

最后，通过勘察设计，提出了对中国城市总体规划"绿带"和"绿楔"持续有效性的疑问，其确立已久，但规划概念仍遭受质疑。通过深入探究那些普遍被接受、尽管仍不明确的空间规定惯例，可清晰地认识到，应该引入、改进和发展更现代、局部有差异、区域可接受的复杂的景观发展理念。此外，由于中国正处于城市化高潮，城市边缘交界处可能有非城市使用和打造景观的条件。这样做，中国值得称赞的、长期的、现代风格的和非对立的城乡概念，有望回归和谐的当代生活。

彼得·罗为哈佛大学设计研究院建筑与城市规划教授、哈佛大学终身教授。

Eco-City Taihu

Ryder Pearce
Mark Upton
Troy Vaughn

Sitting squarely within one of the "green wedges" envisioned by the 2004 Beijing Master Plan, Taihu Township exemplifies a number of critical, interrelated challenges faced by the Chinese capital as it attempts to accommodate unprecedented urban growth while addressing ecological concerns that are regional in scale. In particular, Beijing's seemingly intractable problem of water scarcity requires fresh thinking about how future metropolitan growth might be tied to aggressive conservation and remediation measures. Our proposal for Taihu demonstrates a holistic strategy based on smart growth and environmental performance.

Lying at the lowest elevation within metropolitan Beijing, Taihu is a natural "sink", burdened with alleviating much of the pollution created elsewhere within the capital region. Our proposal introduces a network of hydrological infrastructure and energy-producing facilities that, at a local level, could also become prototypes for other communities to implement. Large, inviolate ecological buffers would provide defensible edges and encourage higher density as a response to population growth that does not require consumption of more land area. Such measures would help Taihu become a true "eco-city", a prototype from which the rest of the world might learn.

In the township's future, we envision high-density residential areas combined with an advanced agricultural and commercial infrastructure that distributes goods and services equitably to its citizens. Key to implementing our proposal is the notion that Taihu's existing village collectives should become involved in the area's critical planning decisions, maintaining their presence within the community as stewards of the land. This hands-on approach foresees a planned densification of existing villages as well as options to invest in village expansion areas. This scenario creates an economy-of-scale that is largely missing in today's Beijing, where the status quo development approach simply houses opportunities for economic investment rather than creating them.

台湖生态城

赖德 · 皮尔斯
马克 · 厄普顿
特洛伊 · 沃恩

在2004年版的北京市城市总体规划中，台湖镇正好位于某一规划的"绿楔"之中，当首都北京试图在适应空前的城市发展的同时加强区域整体的生态关注时，台湖镇证明了北京所面对的这一系列关键而相互关联的挑战。特别是从表面上看，北京的缺水问题需要有关城市未来发展的全新思维，实际上，这一问题关系到激进保护和弥补的措施。我们的方案建议在精明增长和环境利益的基础上制定整体战略。

台湖镇位于北京市域范围的海拔最低点，是天然的"洼地"，首都区域内其他地区产生的污染在此得以稀释。我们的方案引入了水文基础设施的网络和产生能源的设备，这一设备在某种当地层面上能成为其他社区执行的原型。巨大的生态缓冲区将提供防护性的边界，并鼓励以更高的密度来适应人口增长，以避免消耗更多土地。这些措施将有助于使台湖成为一个真正的"生态城市"，一个可供世界上其他地区借鉴的样板。

在未来的小镇里，我们设想将高密度的住宅区与先进的农业、商业基础设施结合起来，这些基础设施为居民平等地提供商品和服务。实施方案的关键是台湖现有的农村集体应该参与到该地区的关键规划决策中，像社区内土地的管家一样维护这些规划。这种亲自参与决策的方法能使现有村庄的规划更有预见性，亦能预见到在该地区的投资决策。这种情况下创造的规模经济如今在北京几乎消失了，现在的方法仅仅是回避投资机会，而不是创造规模经济。

The establishment of a high-speed rail station – to be shared with Yizhuang's new skyscraper district to the south – represents a tremendous development opportunity for Taihu. The simple gesture of bridging over Beijing-Tianjin Expressway is all that is needed for local residents to gain access to a vast regional transit network. A new satellite urban core is anticipated to emerge in the area north of the station. We propose that a dense new street network and urban block structure should be superimposed here to encourage a high-density mix of uses and to facilitate pedestrian and light-rail connectivity. Parks and open spaces, as well as preserved portions of existing villages, are woven into the fabric of the new district plan.

拔地而起的高铁站，台湖南部与亦庄的新摩天大楼区相邻——预示着台湖巨大的发展机遇。京津高速满足了当地居民融入巨大的区域交通网络的需要。火车站以北将出现一个新的卫星城的中心区。我们建议，应叠加密集的新街道网络和城市街区结构，鼓励高密度混合使用以促进步行和轻轨的交通连接。公园、开放空间和被保留的部分现存乡村，构成了新区规划的基础。

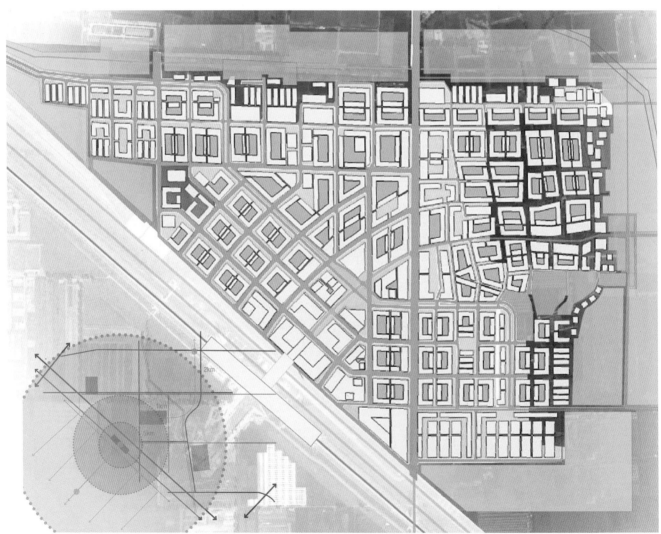

STRUCTURE PLAN   结构规划

NEW TOWN COMMERCIAL CENTER　　　　　　　　　　　　　　　　　　　　　　　　　　　　　　　新城商业中心

MULTI-MODAL TRANSPORTATION HUB　　　　　　　　　　　　　　　　　　　　　　　　　　　　多模式交通枢纽

In our study, Taihu's surface water is treated to provide safe irrigation for the farmland and water for the region's inhabitants. All of the water used to irrigate Taihu's agricultural land enters at a single location near its northwest boundary; today this water arrives highly polluted with organic and industrial waste from sources upstream. Our proposal's first intervention is to locate a series of settlement ponds to remove heavy metals at the point where water flows into Taihu. Water is then directed into the existing canals and throughout the cropland, where remaining organic solids in the water can nourish plant growth. A series of treatment wetlands are introduced at the southern (downstream) end of each irrigated field as a kind of living infrastructure. As water courses through the wetlands, remaining organic waste is removed through phytoremediation. The clean water flowing out can then be diverted into basins to percolate into underlying soils and recharge severely depleted aquifers, or it can be collected in cisterns to provide potable water for adjacent villages.

As part of a regional strategy to reduce the burden of pollution flowing out from Beijing, we propose to de-channelize the Liangshui River to the southeast of Taihu. The expanded waterway's vegetation will filter and cleanse its water. Throughout the seasons, this new dynamic landscape will showcase the natural processes of the river ecology with pockets of wetlands and riparian forest, while the diverse man-made topography will create new recreational and scenic amenities for the people of Taihu and the greater Beijing region.

在我们的研究中，经过处理的台湖地表水为农田灌溉和本地区的居民用水提供了安全保障。用于农业灌溉的水源从台湖镇西北边界流入，如今，这一水源携带来自于上游的有机物和工业废物，已经严重污染。方案中的初次水处理过程是在水体汇入台湖的地点设置一系列沉淀池以去除重金属污染。然后将水体引入现有的运河中，穿过农田，利用水中的有机物滋养作物生长。在每块灌溉土地的南端（河流下游）引入一系列处理过的湿地作为生态基础设施。当水流经湿地时，通过植物修复技术去除剩余的有机物。流出的干净的水可以转移到水池中，渗透到底层土壤和补给严重枯竭的含水层，或者可以收集在贮水槽中为附近村庄提供饮用水。

作为减轻北京流域污染负荷的区域战略的一部分，我们提出将台湖东南部的凉水河去渠化。河道中延伸的植被将水体过滤和净化。岁月流转，这一新的动态景观将与成片的湿地、河岸林一同展示河流生态的自然过程，而多样的人造地形地貌，将为台湖和泛北京区域的居民创造新的休憩景观设施。

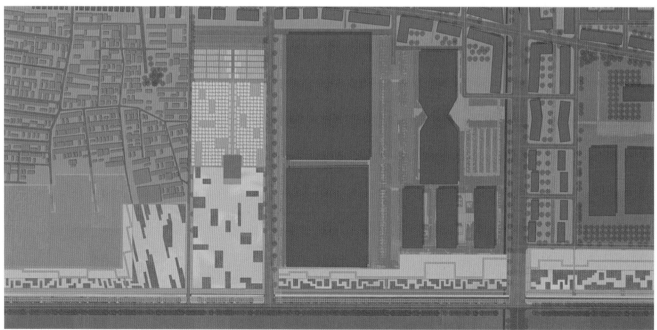

NATURAL SYSTEMS INTERFACE WITH SETTLEMENT

自然和人居结合

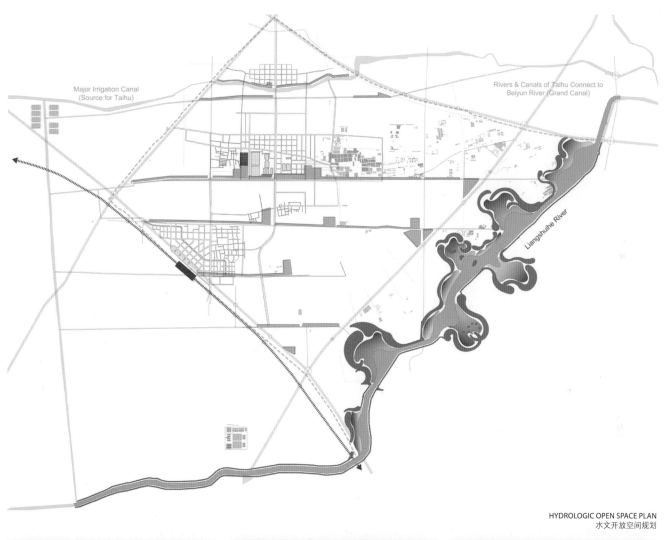

HYDROLOGIC OPEN SPACE PLAN
水文开放空间规划

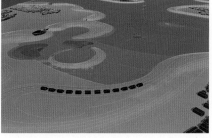

RIVER AND ENVIRONS - JANUARY　　　河流与环境——一月　　　JULY　　　七月　　　OCTOBER　　　十月

In our proposal, the majority of Taihu's land area is neither exclusively urban nor preserved green space. Instead, we aim for an interweaving (or, at times, overlapping) of agricultural land, wetlands, wind and solar energy production, linear parks, transit corridors, and urban development of various scales across the 36 sq-km site. Taihu's identity – as an ecologically progressive place well connected to its agricultural past – will be formed by the myriad instances where built fabric and landscape systems converge to produce characteristic public space.

我们建议台湖土地的主要区域既不用作专门的城市用地，也不作为保留的绿地。相反，我们的目标是将农用地、湿地、风能和太阳能生产用地、线性公园、交通走廊以及横跨36平方公里的场地上的各种尺度的城市发展在空间上交织或时间上重叠在一起。台湖的认同感——作为一个连接过去农业生产的生态先行区——将由无数小区域组成，这些区域把建成机理和景观系统聚合在一起以产生富于特色的开放空间。

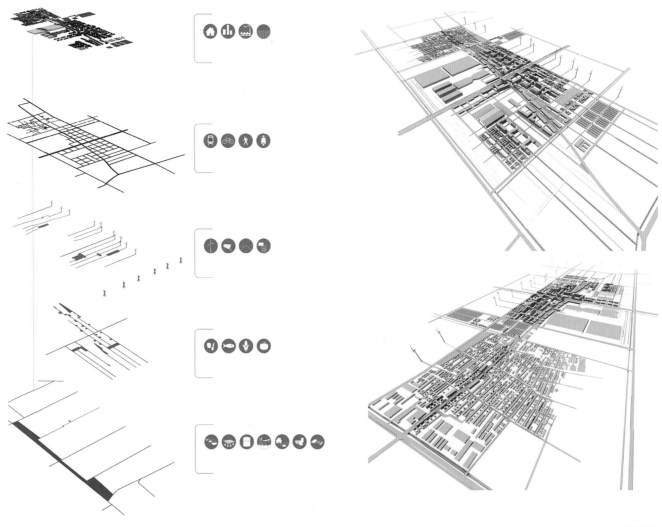

PROGRAMMATIC NETWORKS

规划网络

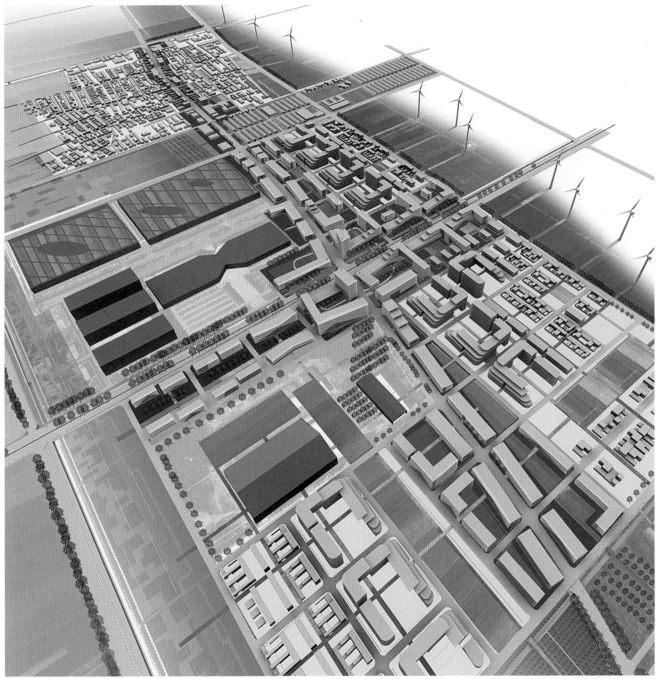

MIXED USE URBAN DEVELOPMENT 城市发展的混合使用

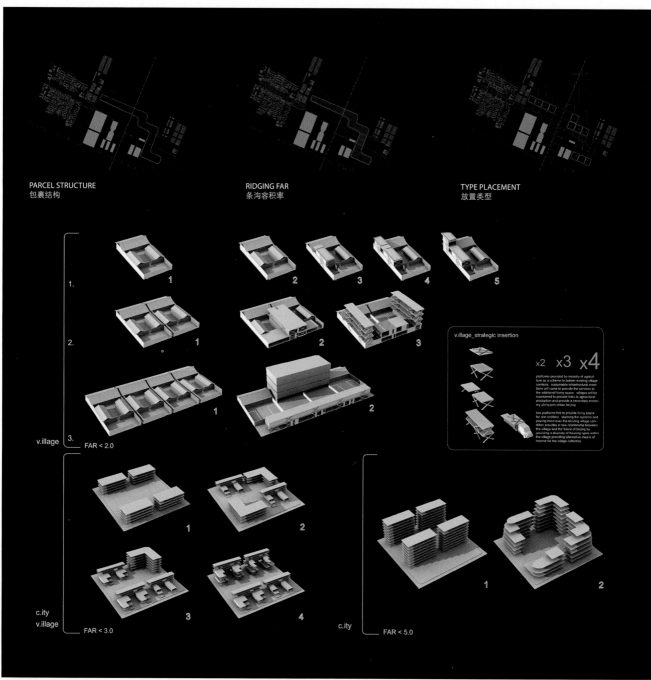

HOUSING TYPOLOGY PROPOSAL 住宅类型方案

URBAN WETLAND SYSTEM 城市湿地系统

NEW TOWN ECOLOGICAL BOULEVARD 新的城市生态休闲道路

Green Network, Green Web

Yenlin Cheng
Hana Disch
Aditya Sawant

By virtue of its geographic position between the two rapidly developing new towns of Tongzhou and Yizhuang and as a stop along the high-speed rail connecting Beijing and Tianjin, Taihu Township is on track to become a major transportation interchange in the near future. Its foreseen role as a transit node has imposed immense urbanization pressure on the area, in apparent conflict with its designation in the Beijing Master Plan as a part of a "green wedge". Rather than accepting a weakly defined "green" designation for land uses across the entirety of the township – only to witness its gradual erosion due to inexorable demands of urbanization – we propose to address the Master Plan's ecological aspirations by investing in a more defensible form of landscape infrastructure: the "green network" or "green web". Composed of interconnected landscape systems such as existing canals and wetlands, windbreaks, woodlands, parks, and other protected open spaces – many of these requiring only modest dimension to be effective – the green network we propose for Taihu, together with new public transit lines, becomes a primary armature for future development rather than remainder or afterthought. Riparian buffer zones are established along the Liangshui River, and the state-owned swaths of land beneath existing power lines are converted into ecological corridors, promoting rich habitat for local fauna and flora. While promoting biodiversity, combating pollution, and otherwise enhancing the region's ecological health, the green web also serves as a civic recreational amenity, accessible to local residents and citizens of Beijing alike. We propose two options for future light-rail lines within the study area – one passing through the heart of Taihu, the other deliberately avoiding it – to demonstrate our overall concept's flexibility.

RENDERING OF A CANAL

绿色网络

陈燕霖
哈娜 · 迪施
阿迪亚 · 萨瓦特

台湖镇位于两个迅速发展的新城——通州和亦庄之间，同时拥有京津城际铁路的车站，依托于其地理位置优势，台湖镇在不久的将来将成为一个主要的交通交换枢纽。台湖镇作为交通节点的功能是可以预见的，这一功能为该地区带来了巨大的城市化压力，很明显，这将与其在北京市总体规划中所承担的"绿楔"功能产生冲突。与其在整体镇域范围的土地利用中附加一个"绿色"的概念——这只会导致台湖镇在城市化进程的强烈需求下土地逐渐侵蚀——我们建议通过建设更具防护性的景观基础设施，如绿栅或绿网，实现总体规划的生态愿景。我们建议的台湖镇的绿网将同新的公共交通线路一起，成为未来发展，而不是残余或事后弥补的"基础枢纽"。而这个绿网将由相互连接的景观系统组成，比如现有的运河和湿地、防风林、林地、公园，和其他防护的开放空间——许多只需适当的尺度便可有效。沿凉水河河岸设置河道缓冲区，将现有的输电线路下大片的土地变成生态走廊，将为乡土动植物群落带来大量的栖息地。绿网将在促进地区生态健康的同时提高生物多样性，防治污染，作为市民休闲便利设施，面向本地开放。我们为研究区域内未来的轻轨线提出两套方案——其一是经过台湖镇的中心，其二是有意绕开中心——以展示我们整体概念的灵活性。

水渠景观意向图

THE MASTER PLAN OF GREEN WEB
绿色网络总平面图

This scheme foresees improved transit connectivity for Taihu in the form of a new regional train line connecting Yizhuang's existing light-rail and high-speed rail to Beijing's airport, as well the extension of Beijing's M1 subway line from Tongzhou to the northern edge of Taihu. Dedicated bus lanes, bicycle and pedestrian paths, rows of shade trees, and other green spaces are added to Taihu's existing main streets and canals to create a durable and richly layered network of public space.

这项计划预计将以新的区域铁路线的方式增强台湖的运输连通性，这一铁路线把亦庄现有的轻轨线和到北京机场的机场快轨连接在一起，同时将北京地铁 M1 线延伸至台湖镇的北缘。公交车专用车道、自行车和步行道、林荫道和其他绿色空间将会添加到台湖的现有主干道路和河流系统中，以创造持久、丰富的开放空间分层网络。

TRANSPORTATION PROPOSAL
交通运输方案

SITE SECTION

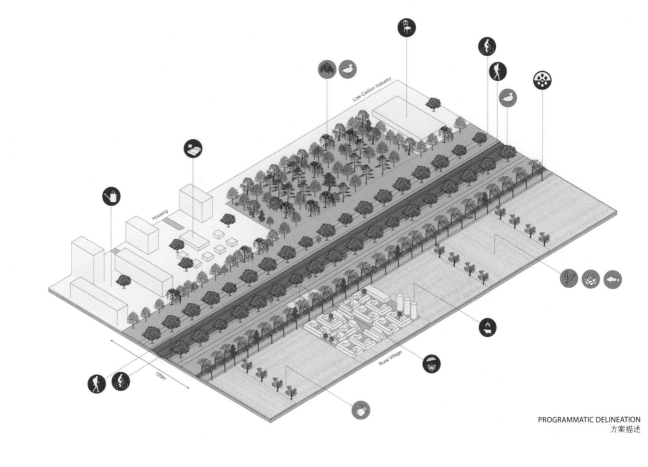

PROGRAMMATIC DELINEATION
方案描述

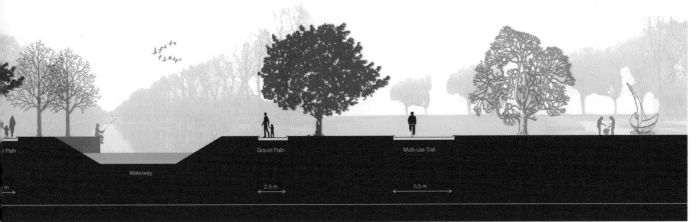

场地剖面

Our demographic analysis of the township reveals a complex set of constituencies, including farmers and businessmen, regional and local officials, members of village collectives, unregistered workers ("floating population"), and so on, all expressing divergent needs and goals for the future of Taihu. We propose that within the framework of an overall "green web" plan, different portions of the site should be allowed to develop in different ways – for example, the areas bordering Yizhuang to the south and Tongzhou to the north will be expected to become more densely urbanized, the rich agricultural lands along the Liangshui River less so – with various forms of landscape corridor serving both as buffer and linkage between them.

我们的乡镇人口分析揭示了复杂的区域构成，包括农民和商人、本地和地方官员、村集体、无本地户籍的工人（流动人口）等等，所有人都表达了对台湖未来的不同需求和目标。我们提出，在一个整体的"绿色网络"规划框架内，研究区域的不同部分，应允许以不同的方式发展——例如南部与亦庄接壤和北部与通州接壤的地区城市化密度有望更高，沿凉水河农耕土地丰富的地区城市化密度则可减少，通过不同形式的景观走廊作为两者之间的缓冲和联系。

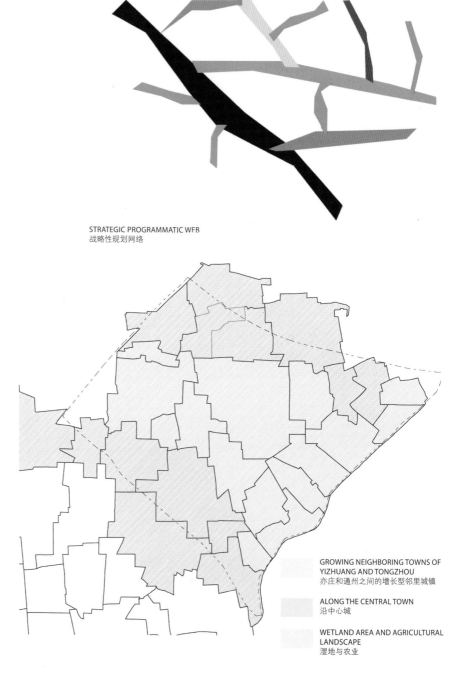

STRATEGIC PROGRAMMATIC WEB
战略性规划网络

GROWING NEIGHBORING TOWNS OF YIZHUANG AND TONGZHOU
亦庄和通州之间的增长型邻里城镇

ALONG THE CENTRAL TOWN
沿中心城

WETLAND AREA AND AGRICULTURAL LANDSCAPE
湿地与农业

URBAN
城市

GREENWAY CIRCULATION SYSTEMS
绿环系统

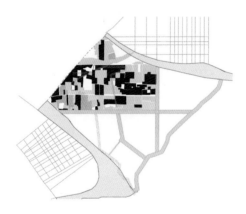

SEMI-URBAN
半城市化

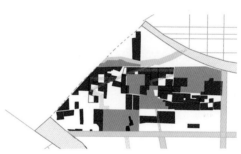

CONSOLIDATION OF OPEN LANDS
合并开放土地

RURAL
乡村

AGRO-TOURISM CORRIDOR AS CONNECTOR
作为连接的乡村游憩廊道

Taihu has relatively abundant natural resources, including fields, orchards, woodlands, as well as a substantial hydrological network composed of the Liangshui River, canals, ditches, lotus and fish ponds, and other man-made wetlands, providing a relatively rich natural substrate to support biodiversity. The continuous growth of industrial development and urban sprawl in this area not only displaces Taihu's essential farmlands but also threatens the reduction and fragmentation of these natural habitats. This proposal expands Taihu's existing canal system so that, in addition to the utilitarian demands of cleansing and providing water for irrigation, larger canals might also serve as a form of scenic structure – or, linked to mass transit systems, as a kind of organizing device for urban growth. The aim is to sponsor a form of urban growth intimately connected with greenways.

In this scheme, a new light rail transit line connecting Yizhuang and the high-speed rail station to the south with Taihu's current administrative center and Tongzhou New Town to the north aligns itself with portions of Taihu's green network to produce a representative civic space. New urban development is encouraged along this hybrid "eco-transit" corridor to form a compact linear city, where most new residents will live within a ten-minute walk of basic services. This dense form of urbanization preserves both agricultural land and residents' livelihood, allowing a gradual transition to ecological agricultural industries of the future. The eco-transit armature bisects two other green corridors within the green web and allows diverse interaction with the landscape, generating both a strong sense of townscape and a strong sense of local cultural identity.

台湖有相对丰富的自然资源，包括田野、果园、林地，以及由凉水河、运河、沟渠、荷塘和鱼塘，以及其他人工湿地所构成的大量水网，这些提供了生物多样性所需的相对丰富的天然本底。这一区域内持续的工业发展与城市扩张不仅取代了台湖镇的基本农田，而且还对这些自然栖息地造成缩小和碎片化的威胁。这一建议扩展了台湖现有的运河系统，除了净化和提供灌溉用水的功能要求，更宽的河道也是作为一种景观结构形式或是作为城市发展一种组织策略与大规模运输系统相连接。其目的是提供一种与绿道紧密联系的城市发展模式。

在此方案中，一条新的轻轨线将亦庄、南部的位于台湖目前行政中心的地铁站和北部的通州新城连接到一起，这条轻轨线与台湖绿色网络相接，构建了一个典型的市民空间。鼓励沿着这种混合的"生态交通"走廊进行新的城市发展，以形成一个紧凑的线性城市，大部分新居民所需的基本服务在10分钟的步行路程以内可以获得。这种城市化高密度形式既保留了农业用地又有市民生活，从而逐步过渡到未来的生态农业产业。这种生态-交通枢纽将绿网中的其余两个绿色走廊一分为二，并与景观发生各种各样的相互影响，形成强烈的城镇景观和本地文化认同感。

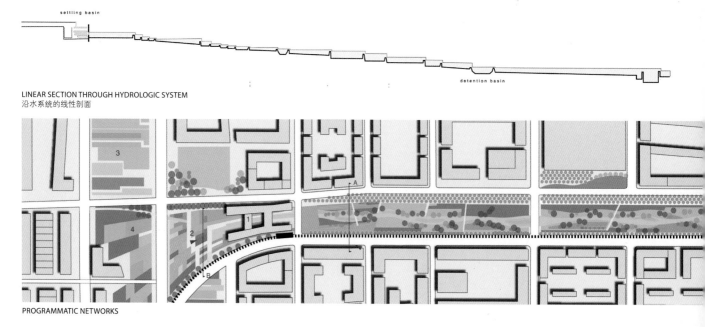

LINEAR SECTION THROUGH HYDROLOGIC SYSTEM
沿水系统的线性剖面

PROGRAMMATIC NETWORKS

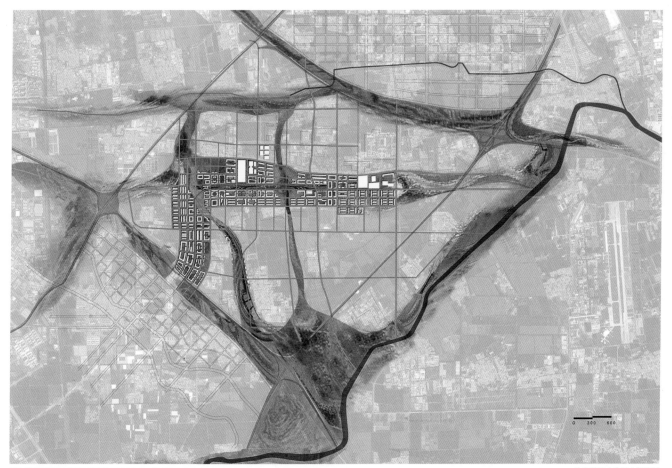

URBAN SETTLEMENT WITHIN ECOLOGICAL FRAMEWORK

生态框架内的城市居住区

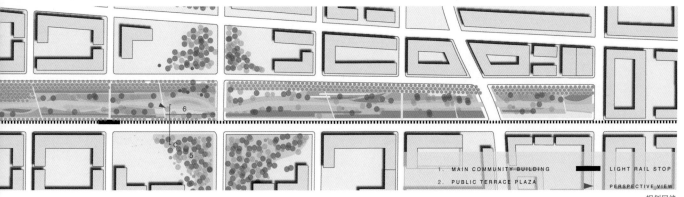

1. MAIN COMMUNITY BUILDING
2. PUBLIC TERRACE PLAZA

LIGHT RAIL STOP

PERSPECTIVE VIEW

规划网络

MULTI-MODAL TRANSPORTATION HUB　　　多模式交通枢纽

URBAN WATER RECHARGE　　　补给城市用水

NATURAL FILTRATION COUPLED WITH DEVELOPMENT 　　　　　　　　　　　　　　　自然过滤结合发展

ENHANCEMENT OF PERI-URBAN CANOPY 　　　　　　　　　　　　　　　增加城郊地区的林荫

Urban Landscape Connectivity

Darin Mano
Hyung Seung Min
Carrie Nelson

城市景观连通性

达林 · 马诺
闵玄胜
卡莉 · 尼尔森

Tongzhou and Yizhuang, two newly planned satellite towns emerging in the southeast periphery of Metropolitan Beijing, exemplify a kind of "instant city" growth that, bleak as it may appear, is hardly surprising given the economic forces at work in today's rapidly urbanizing China. Sandwiched between these two instant cities, the township of Taihu may be fortunate enough to chart a somewhat different development course in the near future, thanks to its designation in the 2004 Beijing Master Plan as one of the radial "green wedges" surrounding the city center. In broad terms, the Master Plan suggests that Taihu's role in serving the region's ecological needs – largely by preserving open green space – might allow it to avoid being swallowed up by typical urban sprawl. But the situation on the ground in Taihu – the gradual transformation of villages and farmland into high-rise housing, industrial, and other uses, together with the soon-to-open high-speed rail station at the township's southern edge – suggests that more targeted and nuanced strategies are needed to realize the potential benefits of Taihu's unique situation.

Our scheme proposes a series of well-defined development zones of different characters to address the complex set of demands emerging in Taihu. Among these zones – which traverse the site in bands, southwest-to-northeast – are swaths of land set aside primarily for agricultural uses; a zone along the Liangshui River reserved for the combined uses of environmental remediation and eco-tourism; and a transit-oriented urban development zone roughly paralleling a proposed light-rail line that connects the centers of Tongzhou and Yizhuang. Within the urban development zone, diverse programmatic uses, densities, housing types, and streetscapes are encouraged to develop through a combination of form-based zoning regulations and rules linking development rights to public space amenity. High-rise development is envisioned for the area surrounding the new high-speed rail station, becoming an economic engine for Taihu.

通州和亦庄，位于北京都市区东南边缘的两个规划中的新卫星城，体现了一种"即时城市"的增长，或许它们也会变得暗淡，但作为推动当今中国快速城市化的经济力量，并不令人惊奇。在2004版的北京市总体规划中，台湖镇被作为中心城外围放射状的"绿楔"之一，得益于此，夹在两个"即时城市"之间的台湖镇或许足够幸运，可以在不久的将来走出某种不同的发展路径。在宏观角度上，总体规划指明了台湖在区域生态需求中的作用——即最大程度地保护开放绿色空间——这避免了台湖在常见的城市蔓延中被蚕食。但台湖土地利用的形式由村庄、农用地向高层住宅、工业和其他用途的转变及台湖南缘高铁站的不日开放表明需要更有针对性和细致的战略将台湖独特状况的潜在益处加以实现。

我们的方案提出了一系列特色不同、定义明确的发展区域以满足台湖出现的多种需求。这些从西南到东北横贯场地的区域，包括大片的农耕土地；包括沿凉水河为环境修复和生态旅游的复合用途所预留的土地；包括交通导向的城市发展用地，这一用地与规划中的轻轨线大致平行，线路连接了通州和亦庄的中心。在城市发展区域内，依据基于区划的法规和关于舒适公共空间的发展权制度，鼓励多种方案的使用、集约化、住宅多样化和街道景观。区域内围绕新的高铁站的高速发展将成为台湖镇的经济引擎。

We foresee that much of Taihu's future growth will be clustered around the high-speed rail station at Yizhuang and therefore propose a set of rules for structuring the area's growth to create a vibrant urban district. Key components of the district plan include 1) the submersion of the highway Beijing-Tianjin highway to improve pedestrian and vehicular access between Taihu and Yizhuang; 2) a mixed use, multi-block development bridging across the station area and including offices, condominiums, and an iconic hotel tower; 3) the extension of scenic canals and linear parks into the urbanized area; 4) a precise set of rules governing building heights and massing to encourage lively and diverse urban environments to emerge; and 5) the provision of landscape buffer zones to north, east, and south to contain sprawl.

我们预见，台湖未来的增长将聚集在地铁亦庄火车站周围，因此提出了一套区域发展的标准，创造充满活力的区域。区域规划的关键组成部分包括：（1）下穿京津高速公路，改善行人及车辆在台湖与亦庄之间的往来；（2）混合使用的多街区发展，衔接车站地区，包括写字楼、公寓和标志性酒店高楼；（3）将景色优美的运河和线性公园延伸到城区；（4）一套精确控制建筑物高度及体量的标准，促进有活力的、多元化城市环境的出现；（5）限定北部、东部和南部的景观缓冲带，控制城市蔓延。

STRUCTURE PLAN - TAIHU , YIZHUANG   结构规划平面图 - 台湖，亦庄

PROGRAMMATIC STATION SECTION   规划车站剖面

SKYLINE VIEW OF TAIHU, YIZHUANG

MASSING - PROGRAM 概念方案图

台湖、亦庄的天际线

In addition to the new high-density development surrounding the high-speed rail station, two further mixed-use districts are proposed in Taihu, each centered on a new light-rail station and incorporating existing institutional and commercial buildings where feasible. Preliminary ideas for built form are based on precedent studies of appropriate low-, medium- and high-rise housing typologies. A comparison of floor-area-ratios (FAR) of existing courtyard houses with those of other examples such as rowhouses, multifamily "walk-up" units (up to five stories), and taller apartment slabs and towers, suggests that far greater density – and better and more diverse living conditions – could be supported in Taihu without sacrificing productive agricultural land or people's connection to greenery.

除围绕高速铁路站的高密度发展外，还在台湖提出了另外两个混合使用区，每个区域都在可行的情况下以一个新的轻轨车站为中心，并纳入现有的公共和商业建筑。建筑形式的初步构想基于对低、中、高层住宅类型的范例研究。比较现有的四合院及其他类型住宅诸如联排别墅、多家庭住宅楼（最多 5 层）、较高的公寓楼等的容积率（FAR），在不牺牲多产农业用地及人们与绿色环境之间的联系下，支撑更大密度、更好、更多样化的生活条件。

STRUCTURE PLAN

结构规划

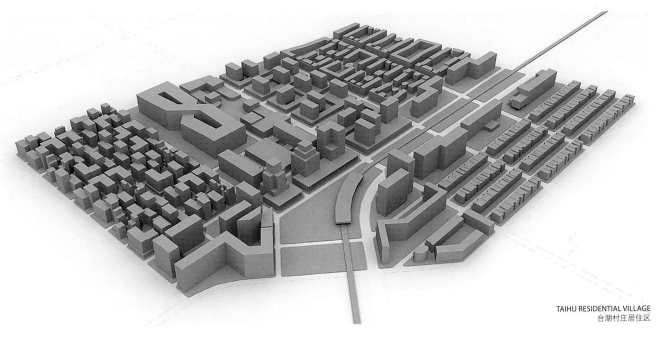

TAIHU RESIDENTIAL VILLAGE
台湖村庄居住区

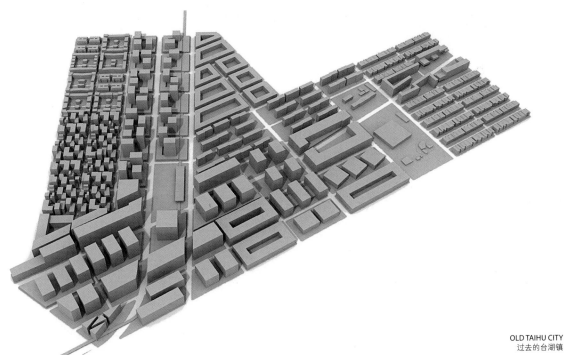

OLD TAIHU CITY
过去的台湖镇

The Liangshui River has its headwaters in the mountains to the north and west of Beijing; it traverses the heart of the city before crossing into the Taihu area, moving south toward Tianjin, and flowing out into Bohai Bay. The river travels through dense urban areas as well as agricultural and industrial areas; urban stormwater, agricultural runoff high in nutrients, untreated sewage from villages, and industrial wastewater mark the inputs to the river. With high levels of pollution and the critical shortages of water that already exist within the region as Beijing rapidly depletes its aquifers (its largest source of fresh water), the need for robust water treatment and recycling and for groundwater recharge becomes critical. The sparsely populated portion of Taihu that lies between the Sixth Ring Road and the Liangshui River offers an exceptional opportunity to create a large-scale system of treatment wetlands that could clean more than 3% of the river's total volume, while creating new opportunities for agricultural production, economic development, and cultural and environmental education on the site.

The proposed Taihu Wetlands Park creates multiple layers of program to derive maximum benefit from this treatment system, including the production of fish for food, habitat for migrating birds, and spaces for local tourism integrated into the landscape. Existing villages are preserved, and each develops a different function and identity within the park; existing buildings are reprogrammed for use as workspaces for artists, restaurant and market stalls, conference spaces, and tourist resorts. Villages within the park are linked together through new bicycle paths traversing the wetlands, while vehicular traffic is restricted to existing primary roads. The wetland areas connect back to Taihu's urbanized nodes through green corridors that utilize existing rights-of-way along canals and power lines.

凉水河的源头在北京北部和西部山区；它穿越北京市中心之后流入台湖地区，向南流向天津，然后汇入渤海湾。这条河穿过密集的市区以及农业和工业区；城市雨水、农业富营养化排水、村庄未经处理的污水及工业废水都排入河中。严重污染及水资源紧缺迅速消耗了北京的含水层（其为最大的淡水来源），有效水处理和循环利用以及地下水补给的需求十分迫切。台湖位于六环路与凉水河之间的人口稀少地段，凉水河提供了不可多得的、创造大型湿地污水处理系统的机会，可以净化3%以上的河流总水量，同时为场地的农业生产、经济发展和文化环境教育创造新机遇。

所提出的台湖湿地公园计划创建了多层次的项目，使得其效益最大化，包括鱼类食物生产、候鸟栖息地及融入当地景观的旅游空间等。保留现有的村庄，每个村庄在公园中都有不同的功能及个性；现存建筑被重新规划，用作艺术家的工作室、餐厅、市场摊位、会议场所、旅游度假区。公园内的村庄用新的穿越湿地的自行车道连接起来，而机动车辆交通被限定在现有主要道路。湿地区域利用现有沿沟渠的公共线路和电线建立绿色廊道，连接了台湖的城市节点。

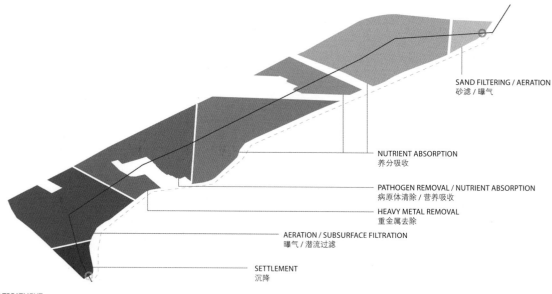

SAND FILTERING / AERATION
砂滤 / 曝气

NUTRIENT ABSORPTION
养分吸收

PATHOGEN REMOVAL / NUTRIENT ABSORPTION
病原体清除 / 营养吸收

HEAVY METAL REMOVAL
重金属去除

AERATION / SUBSURFACE FILTRATION
曝气 / 潜流过滤

SETTLEMENT
沉降

WATER TREATMENT
水处理

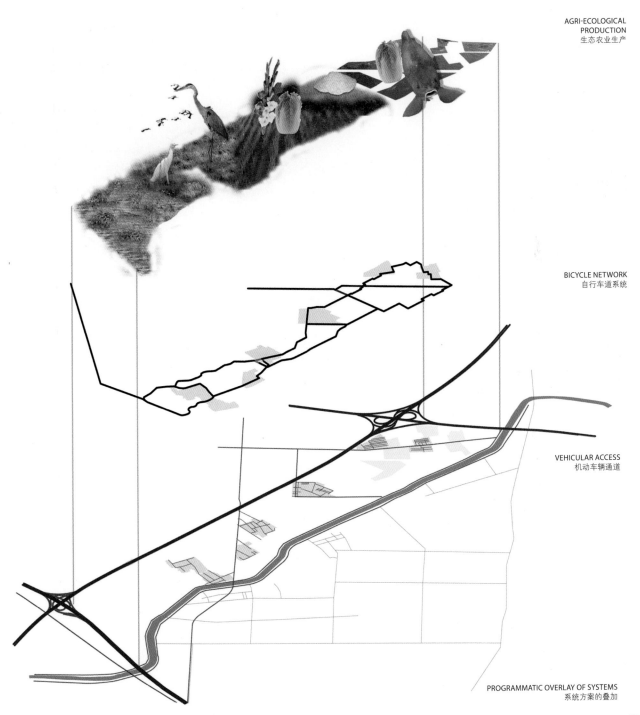

AGRI-ECOLOGICAL PRODUCTION
生态农业生产

BICYCLE NETWORK
自行车道系统

VEHICULAR ACCESS
机动车辆通道

PROGRAMMATIC OVERLAY OF SYSTEMS
系统方案的叠加

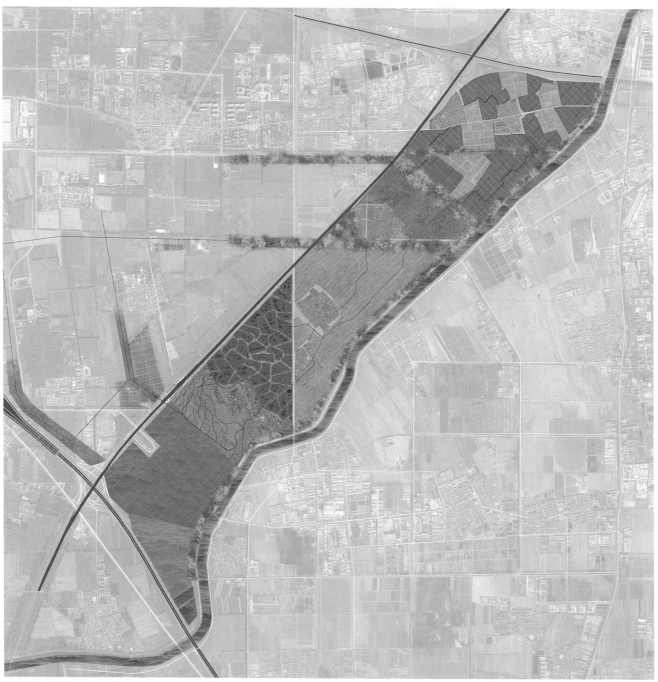

ECOLOGICAL STRUCTURE PLAN

生态结构规划

RECREATION FORMED BY NATURAL SYSTEMS　　　　　　　　　　　　　　　　　　　　　　　　自然形成的游憩空间

NATURAL SYSTEMS AS AESTHETIC AND CIVIC AMENITY　　　　　　　　　　　　　　　　　赏心悦目的自然生态

Nested Scales of Urbanization

Cameron Barradale
Fadi Masoud
Andreas Viglakis

城市化的叠加尺度

卡梅伦 · 巴拉戴
法蒂 · 马苏德
安德烈亚斯 · 威格拉基

The Taihu site sits in the crosshairs of the two largest wheat-producing regions of China, as well as within the increasingly urbanized corridor between Beijing, China's administrative center, and Bohai, its associated rapidly industrializing coastal area. The position of the Taihu region at this intersection of large-scale urban and agricultural systems imbues it with crucial strategic importance to the sustainability of northern China. To create a catalyst for appropriate regional and local growth, we propose for Taihu Township an economy based on strong agricultural and industrial connections. This will be a new kind of agricultural urbanism: one focused on research, increased productivity, and the appropriate allocation of social and urban capital. It embodies a new scalable urban typology for peri-urban China writ large.

Regionally, densification of several urban nodes along the Beijing-Tianjin (Beijing-Bohai) corridor seeks to synthesize the need to accommodate an influx of 4.5 million people projected for the next fifteen years, the desire to create open space systems throughout the perimeter of swelling Beijing, and the necessity to secure the little remaining arable land in the region. By greatly densifying urbanized areas, the scheme preserves regional boundaries with growth clustered around appropriately allocated mass transit, infrastructure, and industry.

Within the local Taihu study area, our strategy proposes a nested system of urban and agricultural overlaps based on specific synergies desired within the region. As these systems overlay, they begin to hint at urban form. A lattice is established across the site, the lands within becoming crucial to the performance of the villages, cities, and region as a whole. This systemic scheme becomes the driver of the urban condition and socio-economic reality of the place. Woven together within an efficient transit, open space, and road network formed by the interaction of proposed and existing conditions, a mesh of productivity emerges. Thus, open space and urban boundaries are preserved though leveraging their values as urban assets, creating a reliance of each part on the overall whole.

台湖位于中国两个最大的小麦生产区的交汇处，同时位于北京和快速工业化的环渤海地区所形成的日趋城市化的廊道之间。在大型城市和农业系统交汇处的区位，赋予了台湖对中国北方可持续发展的战略重要性。为了促进区域和地方的适当发展，我们建议台湖应以强大的农业和工业相结合作为经济基础。这将是一种新型农业城市化：它致力于研究、生产力的提高、社会和城市资本的合理配置，体现了中国广大城郊一种新的可推广的城市类型。

从区域上看，沿京津走廊（北京 - 环渤海）几个密集的城市节点，应综合考虑对未来15年预计涌入的450万人口的需求、在北京周边创建开放空间系统的需求及保证区域少量剩余耕地的必要性。该方案通过大幅增加城市化地区的密度，通过合理配置交通、基础设施、产业三者的集聚发展来保留区域边界。

在台湖研究区内，我们基于区域对特定协同作用的需求，提出一种城市与农业叠加的系统策略。这些系统叠加后，开始显现一种城市形态。在此区域建立一种网格结构，土地在村庄、城市和区域一体中的角色变得至关重要。这种系统方案成为了城市环境和社会经济现实的驱动力。方案及现状条件相互作用形成的高效交通、开放空间和道路网络交织在一起，产生一种紧密配合的生产力。因此，通过城市资产价值的杠杆作用，创建一个依赖各部分的综合整体，使开放空间和城市边界得以保持。

COMPUTER RENDERING OF THE NESTED SCALES OF URBANIZATION
城市化的叠加尺度效果图

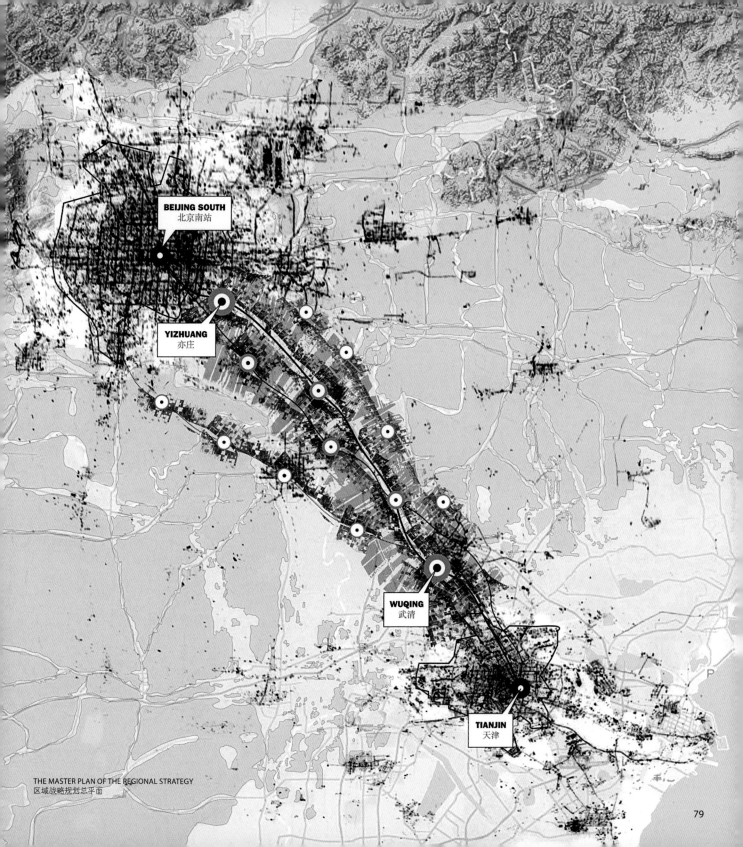

THE MASTER PLAN OF THE REGIONAL STRATEGY
区域战略规划总平面

As the Bohai region's population is expected to increase by 4.5 million over the next 15 years, a strategy which accomodates 75% of the influx within the Beijing-Tianjin corridor is established. The resultant regional settlement pattern runs contrary to those currently employed, seeking to concentrate density in distinct nodes along transit corridors and within productive agricultural zones, rather than allowing the population to either sprawl unchecked into rural areas or preserving open space as static protected lands.

预计未来 15 年内环渤海地区的人口将增加 450 万，因此提出了容纳 75% 涌入京津走廊人口的策略。由此产生的区域聚落形态与现状背道而驰，应试图沿着运输廊道及高效农业区内不同的节点集中人口，而不是允许其无限制扩张到农村地区及受保护的开放空间。

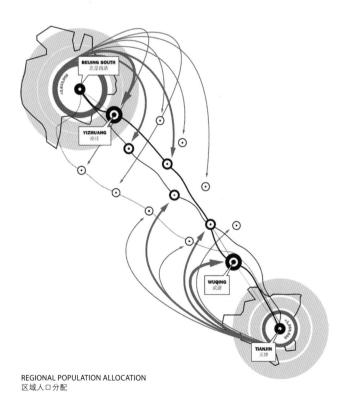

REGIONAL POPULATION ALLOCATION
区域人口分配

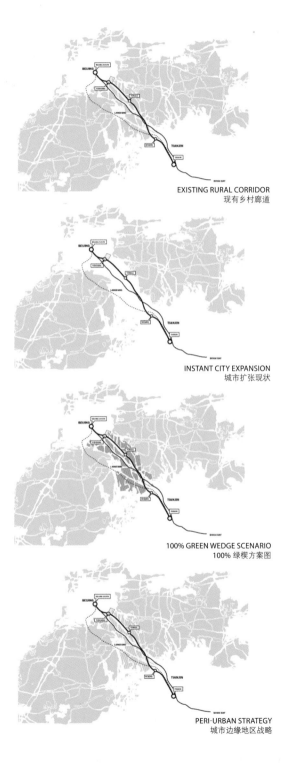

EXISTING RURAL CORRIDOR
现有乡村廊道

INSTANT CITY EXPANSION
城市扩张现状

100% GREEN WEDGE SCENARIO
100% 绿楔方案图

PERI-URBAN STRATEGY
城市边缘地区战略

BOHAI BAY INDUSTRIAL CONCENTRATION   渤海湾地区产业集中度

The New York Times

03/07/11

"VIRTUALLY ALL OF THE WHEAT GROWN IN CHINA IS FAIRLY LOW QUALITY, WHICH WORKS FINE FOR MAKING NOODLES. **CHINA IMPORTS SOME HIGH-GRADE WHEAT EVERY YEAR FOR USE IN BREADS AND PASTRIES, WHICH ARE BECOMING INCREASINGLY POPULAR IN THE CITIES.** CHINESE-GROWN WHEAT IS ALMOST NEVER SUITABLE FOR MAKING CROISSANTS..."

The Washington Post

03/11/11

"HOW DO YOU FEED MORE THAN 1 BILLION PEOPLE? THIS QUESTION VEXES CHINA'S LEADERS, MANY OF WHOM ARE SURVIVORS OF THE GREAT FAMINE, IN WHICH 30 MILLION PEOPLE STARVED TO DEATH BETWEEN 1959 AND 1961. LAST YEAR, **IN AN EFFORT TO HALT RISING FOOD PRICES, THE GOVERNMENT AUCTIONED CORN, WHEAT, RICE AND SOYBEANS FROM STATE RESERVES. AND IN RECENT YEARS, CHINA HAS BROUGHT OR LEASED LAND IN OTHER COUNTRIES** FROM SUDAN TO INDONESIA TO PRODUCE FOOD AND BIOFUELS, BUT THERE IS LITTLE TO SHOW IN PRODUCTION FROM THESE LANDS SO FAR."

EARTH POLICY INSTITUTE

03/23/11

"CHINA'S FOOD SUPPLY IS TIGHTENING. IN NOVEMBER 2010, THE FOOD PRICE INDEX WAS UP A POLITICALLY DANGEROUS 12 PERCENT OVER A YEAR EARLIER. NOW AFTER 15 YEARS OF NEAR SELF-SUFFICIENCY IN GRAIN, IT SEEMS LIKELY THAT **CHINA WILL SOON TURN TO THE WORLD MARKET FOR MASSIVE GRAIN IMPORTS,** AS IT ALREADY HAS DONE FOR 80 PERCENT OF ITS SOYBEANS."

MAJOR WHEAT PRODUCING REGIONS WITH ASSOCIATED DISTRIBUTION   小麦主产区及分布

As the urban fabric begins to emerge, corridors containing the highest intensity of urban diversity become apparent. Further analysis of these corridors reveals specific synergies between urban settlement, agriculture, natural systems, and transportation. These instances of diverse and overlapping programs, occurring most often at the intersection of multiple edges, become hot spots of intense use within the overall plan. Necessity dictates a need to prescribe a more specific urban form for these areas than that of the overall structure plan, as they become central to the functional and aesthetic success of Taihu.

随着城市构造开始出现，多样性最丰富的城市廊道变得明显。对这些廊道的进一步分析揭示了城市居住区、农业、自然系统和交通间具体的协同作用。这些叠加不同方案的例子，最常发生在多种边缘的交叉口，成为总体规划中被高度使用的热点。必须制定一种比城市总体规划更具体的城市形态，这是台湖在功能和美学上取得成功的关键。

PERI-URBAN STRATEGY
城市边缘地区战略

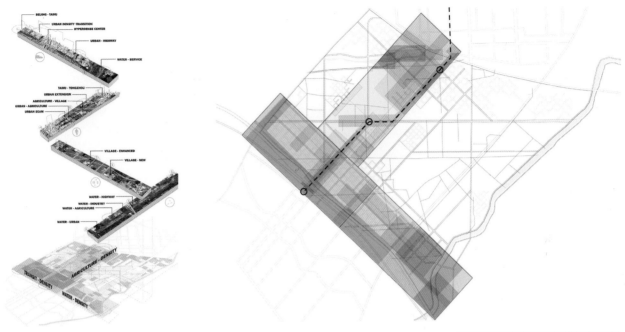

INSTANCES OF ADJACENCY
邻接的实例

STRUCTURING OF PROGRAMMATIC ADJACENCIES
方案邻接的结构

AGRICULTURE - OPEN SPACE - URBAN HOTSPOT

农业 - 开放空间 - 城市热点

The design of three distinct zones establishes models of development that can be selectively broadcast over the entire Taihu site. The Urban-Transport Hub accommodates the high-speed and light rail stations, as well as the highway between Beijing and Tianjin. Pedestrian and vehicular connections between Taihu and Yizhuang are carefully articulated: the interplay between interior and exterior volumes along this corridor unites the transit hub within a dense urban fabric.

The Urban-Hydrological Interface de-channelizes the Liangshui River along the 6th Ring Road to serve as a productive hydrological infrastructure for water collection and treatment, agricultural irrigation, groundwater recharge, and wildlife habitat. Moreover, the resultant wetland and river parks provide scenic recreation along the 6th Ring Road and raise awareness of water's value.

The Urban-Agriculture Interface is nested within new or revitalized villages surrounded by agricultural lands, as well as urban extensions into those lands. Service industries coupled with smaller, high-intensity farms and larger land parcels strike a balance between productive industry, urban settlement, and open space. Agricultural research, educational facilities, and like-minded industries become essential in the creation of a productive urban-open space system that promotes innovation and economic advancement, while also serving as a transition between urban Beijing and adjacent rural lands.

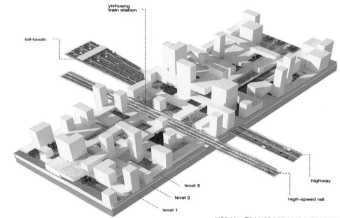

URBAN - TRANSPORT HUB INTERFACE
城市 - 交通枢纽结合

设计建立了三种不同地段的发展模式，可以选择性地在整个台湖镇推广。城市交通枢纽包括高铁、轻轨车站及北京与天津之间的高速公路。应审慎组织台湖、亦庄之间行人与车辆的连接：廊道内外空间的相互作用，将城市密集区内的交通枢纽统一起来。

城市 - 水文的结合开辟了沿六环路的凉水河，使其成为水收集和处理、农业灌溉、地下水补给和野生动物栖息地的高效水文基础设施。此外，由此产生的湿地和河边公园提供了沿六环路的景观游憩，并提高社会各界对水价值的认知。

城市 - 农业的结合位于被耕地包围的新村庄，城市也会扩张到此。与小规模、高效农田和大面积地块相结合的服务业，在生产工业、城市居住区和开放空间之间寻求平衡点。农业科研、教育机构和相关行业对于创造生产性城市开放空间系统十分重要。该系统可以促进创新和经济发展，同时也作为北京市区与郊区之间的过渡。

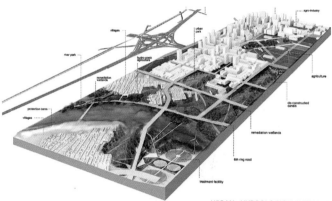

URBAN - HYDROLOGICAL INTERFACE
城市 - 水文结合

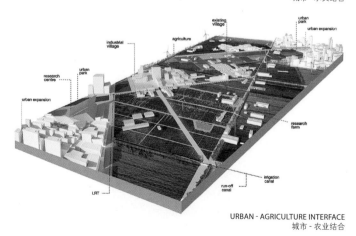

URBAN - AGRICULTURE INTERFACE
城市 - 农业结合

TYPICAL VIEWS CORRESPONDING TO INTERFACES (OPPOSITE)　　　　　　　　　　　　　　　　　　　　　　　　　　　与各种结合相对应的效果图

TAIHU SETTLEMENT

台湖居住区

In the Field: Taihu / Beijing
场地调研：台湖 / 北京

Collaboration / GSD Studio
合作 / 哈佛设计课程

Afterword
by Kongjian Yu
后记
俞孔坚

Charles Waldheim, chair of the department of landscape architecture at the Harvard GSD, was kidding with me the other day: "Each time I expect you to do a smaller studio with a smaller site, but each time you make it even bigger." I was laughing, because this time we have an "even bigger" studio, with the largest instructor team perhaps in the history of GSD studio instruction -- four faculty members combined with a team of diverse expertise in architecture, landscape architecture, urban planning and design, GIS and spatial analysis. The site is over 30 square kilometers. We cannot make it smaller simply because of the unprecedented scale, speed and complexity of urbanization and landscape change in China at this time, particularly in Beijing. The site is the Taihu township, in one of the hottest regions around Beijing.

The challenges faced in Taihu are tremendous: ecological, environmental, cultural, social, technological and economic among others. But what was even more challenging for our students was to understand the rationale behind the unbelievable landscape mosaic of Taihu. If one can understand Taihu, one can also understand Beijing and China. That is the main reason we choose Taihu for the studio. I vividly remember how the mayor of the town defended the development plan of the proposed new town seat, sited in the middle of nowhere, totally away from the more obvious and suitable location around the newly built transit station, simply because it had been decided so. He remained adamant even when he agreed that the proposed site by the GSD students and faculty was much better from social and economic perspectives. Such an irrational position has certainly puzzled our students and faculty alike. But is his position really out of logic, or is it that we, who have been rationally educated by western theories of urbanism and development, simply do not understand the logic behind the decision makers? I tend to believe the latter. A likely scenario is that the selected site was decided before the transit station was built, and the amount of developable land allocated to the specific site has been determined by the municipal government. Any change in the location of the development means a formidable process that may take forever, certainly beyond the administrative term of the current leadership. Immediate development of the land is the key to reaching their GDP goal. Besides, any waiting for a better plan means the loss of the obvious development right. This is just one of many seemingly illogical practice that our GSD students did not understand. Other phenomena that confused our students include: relocating villagers, who once occupied the courtyards in many of the small villages, to skyscrapers far away from their farms, wiping out old villages; grand avenues built through farmland with a dead end; the largest book store in Asia built right in the middle of an agricultural field in the middle of nowhere...etc. It is certainly challenging for planning and design students, who come from totally different social, cultural and political systems, to fully grasp such phenomena and to propose solutions to the landscape that reflect the complicated social formation and values.

To meet these challenges, our students were lucky to have Vice Minister Qiu Baoxin, from the Ministry of Housing and Urban and Rural Develop-

哈佛大学设计学院景观设计学系主任查尔斯·瓦尔德海姆在不久前开玩笑说："每次我都希望你能在一个比较小的场地开展小型课程，但每次你都让它变得更大。"我笑了，因为这次我们的课程"更大"，在建筑、景观设计、城市规划与设计、GIS和空间分析领域不同专长的四组教员，可能是哈佛大学设计学院历史上最庞大的课程指导队伍。场地是北京周边热点区域之一的台湖镇，面积约为30平方公里。我们不能简单地把场地范围缩小，因为当代中国，尤其是北京的城市化和景观变化的规模、速度和复杂性是前所未有的。

台湖镇面临着巨大的挑战：生态、环境、文化、社会、科技和经济等等。但对于我们的学生更具有挑战性的是难以理解台湖嵌合景观背后的原理。如果能理解台湖，也就能理解北京和中国，这是我们的课程选择研究台湖的主要原因。我清楚记得镇长是如何为在完全远离新建交通站，而不选择更明显和合适的地方建立新城的发展规划辩护，仅仅因为这个方案已经确定了。即使他同意哈佛师生提出的场地规划具有更好的社会经济效益，他仍然坚持原有的方案。学生和老师对这个不合理的选址有相似的困惑。但是镇长的观点真的没有逻辑吗？或者是受过西方城市主义理性教育的我们，不了解决策者的逻辑？我倾向于后者。相似的情况是，所选场地在车站建立之前就已有规划，地方政府已经确定了部分具体地块的开发。发展中选址的任何变化意味着这将是一个艰巨的过程，可能超越了现任领导的行政任期。直接开发土地是达到GDP目标的关键。此外，期待一个更好的规划意味着丧失眼前的发展机遇。这是哈佛大学设计学院学生不理解的，看来似乎是不合逻辑的实践之一。让学生感觉困惑的现象还有：推倒场地上的旧村庄，将村民重新安置到远离田地的高楼中；穿过农田的大马路；坐落于农田中央的亚洲最大图书城……我们的规划和设计专业学生来自于完全不同的社会、文化和政治体系，对他们而言，要完全理解这些反映了复杂的社会模式和价值观的景观并提出方案，的确具有挑战性。

学生们有幸听取了住房和城乡建设部副部长仇保兴关于中国城市化进程的讲座及镇长倪德才对台湖镇的愿景，这使他们得以

ment, to lecture on the urbanization process in China, and the Mayor of Taihu Township, Mr. Ni Decai, to share with us his vision. The President of the agricultural enterprise "The United Nation of Tomato" also extended a warm invitation to us to visit and dine in his huge modern green houses, where we could hear the local villagers describe their current living conditions. Faculty and students from Peking University acted both as collaborators and as a bridge between the GSD students and the locals.

The result was encouragingly successful. GSD students gradually understood the dual land ownership system in China, something many of our students never heard about. They now understand the rural and urban Hukou system (Household Registration System) in China, again something totally new to them. They also better understand the complicated land development right system, etc. Alternative solutions were proposed to meet challenges that a complicated peri-urban township is now facing. These solutions might seem to be too idealistic or academic to be implemented. But believe it or not, these unbelievable dream solutions could somehow come to be realized in the very near future - this time, in a positively irrational way. Just one year after our studio's site visit, the Beijing municipal government decided to allocate 60 square kilometers of ecological land to improve the environment of Beijing. And Taihu township, the site of our studio, has been selected as a demonstration site for ecological transformation, partly due to our studio work and the ideas our students presented to the mayor of Taihu. Their proposals–presented in this publication – are now having a positive impact on the town's future.

I am quite sure our next studio will be "even bigger."

应对上述挑战。农业企业"番茄联合国"的总经理也热情地欢迎我们参观大型现代化温室并用餐,当地人给我们讲述了他们的生活现状。北京大学师生既是合作者又充当了哈佛大学设计学院学生与当地人沟通的桥梁。

结果令人鼓舞。哈佛大学设计学院的学生们逐渐了解到中国特色的土地所有制,而很多中国学生都不清楚。他们也了解了中国农村与城镇户口体系及复杂的土地发展权属系统等等。多解规划有望解决当前城郊乡镇的复杂问题。本书中的方案也许看起来太理想化或过于学术而难以实现。但不管你相信与否,这些理想化的解决方案也许会在不久的未来得以实现。就在我们参观场地一年后,北京市政府划定60平方公里的生态用地来改善北京环境。台湖镇成为生态转型示范点,部分归功于我们的课程设计及方案。本书展示的方案正积极推动着台湖镇的发展。

我确信下次课程将会"更大"!

图书在版编目（CIP）数据

台湖展望——中国城乡结合带景观多解规划：北京东南部案例／俞孔坚等编著．
北京：中国建筑工业出版社，2012.8
ISBN 978-7-112-14460-0

Ⅰ.①台… Ⅱ.①俞… Ⅲ.①城市景观－景观规划－案例－北京市 Ⅳ.① TU-856

中国版本图书馆CIP数据核字（2012）第146223号

责任编辑：郑淮兵　杜一鸣
责任校对：张　颖　王雪竹

台湖展望
——中国城乡结合带景观多解规划：北京东南部案例

俞孔坚　彼得·罗　斯蒂芬·欧文　马克·马利根　编著
卡梅伦·巴拉戴　罗伯特·米格尔　编辑

＊

中国建筑工业出版社出版、发行（北京西郊百万庄）
各地新华书店、建筑书店经销
北京嘉泰利德公司制版
北京盛通印刷股份有限公司印刷

＊

开本：889×1194毫米　1/20　印张：$4\frac{4}{5}$　字数：200千字
2012年8月第一版　2012年8月第一次印刷
定价：45.00元
ISBN 978-7-112-14460-0
（22504）

版权所有　翻印必究
如有印装质量问题，可寄本社退换
（邮政编码 100037）